Ethical Digital Technology in Practice

Digital technology is about people. It is about those who plan, develop and implement applications which other people use and are affected by. It is about the impact on all these people as well as on the world at large. **Ethical Digital Technology in Practice** takes a real-world perspective to explore these impacts over time and discover ways in which to promote ethical digital technology through good practice. It draws upon the author's published articles in trade magazines, professional journals and online blogs. These are synthesised into a blueprint which addresses, in a practical manner, the societal issues surrounding the increasing use and abuse of digital technology. It is a follow-up book to the author's book **The Evolving Landscape of Ethical Digital Technology**, which has a researcher's perspective.

This book is a hands-on account of the computer revolution from 1995 to the current day when the world is increasingly dependent on digital technology. It explores some of the social and ethical issues that are part of this revolution. This is not a book about deep philosophical and technical concepts. Nor does it claim to be comprehensive. It is the author's personal account of technological change and its effects on people. It is written by a boy who was smitten by computer technology at the age of 15, became a computer professional and subsequently spent many years showing young people how to develop and use digital technology in a good way. It is a book based upon the author's engagement with practitioners, academics and students during the period as well as his continued fascination with this fantastic technology. **Ethical Digital Technology in Practice** is a book about the real world, of what has happened and what might happen as digital technology continues to pervade.

Ethical Digital Technology in Practice

Simon Rogerson

CRC Press
Taylor & Francis Group
Boca Raton London

CRC Press is an imprint of the
Taylor & Francis Group, an **informa** business

AN AUERBACH BOOK

First edition published 2023
by CRC Press
6000 Broken Sound Parkway NW, Suite 300, Boca Raton, FL 33487–2742

and by CRC Press
4 Park Square, Milton Park, Abingdon, Oxon, OX14 4RN

CRC Press is an imprint of Taylor & Francis Group, LLC

ISBN: 978-1-032-31296-5 (hbk)
ISBN: 978-1-032-14530-3 (pbk)
ISBN: 978-1-003-30907-9 (ebk)

DOI: 10.1201/9781003309079

Typeset in Garamond
by Apex CoVantage, LLC

To my wife Anne

Over 50 years

of

your love, support and friendship

Contents

Foreword

This is an important book. Ethics is not an easy topic, and arguably the ethics of IT is less so—not least due to its potential for developing and evolving in ways that are either unforeseen or unimaginable.

History has shown that moral or ethical scruples rarely put a brake on technological research, development or deployment. Digital technologies have changed the world, as the world, in turn, has changed digital technology. This is a natural part of the development of humanity as a technologically able species; it is a circle, not always a virtuous one, with the inevitable consequence that the cat, once out of the bag, does what cats do. What would have become of humanity if zealous ethical scrutiny had doused the development of fire? Blunted the invention of knives? Vetoed the exploitation of resources? It makes an interesting thought experiment, and no doubt future generations will wonder similarly about the internet, smartphones, social media and self-driving cars. As a species, we have become utterly dependent on this technology, and that has changed us both in ways we can see now, and in ways we currently cannot fathom. Humans, though, have long reached the level of sophistication that allows us to "what if" both ways. We can ask questions; we can model scenarios; we can prepare and plan.

While sometimes unwelcome, applying ethics to IT is not intractable—with appropriate training and guidance, and given appropriate space, IT practitioners can—and do—apply ethical considerations to information systems developments.

CEPIS (the Council of European Professional Informatics Societies), where I chair the *Ethics Special Interest Network*, promotes the development and interests of the IT Profession within European institutions, industry groups, and wider society. It very much echoes Don Gotterbarn's view that "ethics is the conscience of the computing professionals" (Gotterbarn, 2016), and is an ardent supporter of Prof. Rogerson's view, together with Don Gotterbarn and Keith Miller, that "ethical guidance and training must have practical worth if the so-called IT profession is to be recognised as a profession both in name and in deed" (see *Section 5.17 Ethics of the IT Profession*). CEPIS' focus is very much on how ethics can be made practical, with appropriate training, tools and supports, so that it can be made part of the normal working life of the practitioner.

In this second collection of essays and papers, Prof. Rogerson looks squarely at these practical aspects of ethics in Digital Technology—what are the questions, what are the considerations, what rules and principles can be applied, how can considerations be evaluated and decisions made that create value rather than impede?

The book is a journey, encompassing Prof. Rogerson's own travels through the universe of ethics spanning 30 years or so. It is on the one hand a travelogue, documenting Prof Rogerson's own observations as he makes that journey; at the same time, it is a *vade mecum*, a guidebook through the territory of ethics that Prof. Rogerson has explored, and full of practical guidance.

Throughout seven chapters, Prof. Rogerson examines key topics with practical considerations; first setting the scene, and then exploring Data and Information; Systems and Applications; Practice and Method; Regulation and Policy; Practical Considerations; and ending with a Synthesis of all the ideas explored. Each section comes with a rich set of references—a treasure trove in its own right for those wanting to explore further.

My own interests in this area as CEPIS Ethics Chair are well covered:

- The need for comprehensive and structured ethical training, education and practical tools for IT professionals, to facilitate the application of ethical thinking in digital technologies (bookended by "Training for Ethics" and "Rebooting Ethics Education in the Digital Age") is a key theme.
- The importance of the information society as a resource for all (and the potential hazards arising from that) is comprehensively discussed in such sections as "Information for All," "The Data Shadow," "People Issues," "Surveillance," "Social Responsibility," "All-Inclusive Opportunities," "Understanding the Community," "Digital Slavery" and "Is the Digital Divide of the Past, Present or Future?"
- The propensity for "mission creep" and uses beyond intended purposes gets explored in such interesting sections as "The Data Shadow" which uses allegory as a means of asking important questions about the imprint that we leave on the digital world; inferences that can be made, judgements that can be reached
- The challenge of the surveillance society and privacy is explored in "The Data Shadow," "Surveillance," "Facing Up to the Issues," "Data Matching, " and "Waking up to a Surveillance Society."
- Trustworthiness, provenance, and safety of information systems is explored from numerous angles in "Information Provenance" and "Information Integrity in the Information Age" and "Trustworthy Publishing," as well as "Clear and Present Danger" and "Trusting the System" and "Hospital Safety" which ask questions about relying on information systems that have impact in, for example, safety critical circumstances—issues brought home in recent times by tragic deaths involving self-driving cars and aircraft control systems.
- The potential for information systems to create harm is examined in "Risky Business, " "Use and Abuse, " and "Identity Crisis Online, " and (from different perspectives) "Coding Ethics into Technology" and "Ethical Robots".

Interestingly, in Section 1.3 "Ethics Man" Prof Rogerson observes that "ethics are only now being recognised by business as an issue." It is worthwhile considering whether or not (and why) we have progressed much since then. Anecdotally, it remains challenging to raise ethical questions without being viewed as a kind of crank, or to pause a project to examine an ethical issue without standing accused of questionable priorities. And yet there is some real progress driven by increasing need to pay attention to social responsibility, evidenced by the emergence of ethics committees, CEOs (Chief Ethics Officers), corporate codes of conduct, and increasing regulation in certain sectors. Importantly, the European Commission has been showing increasing interest in codes of ethics among IT professionals.

There is no doubt that ethics will play an ever increasing role in digital technologies and their uses, whether internally through their direct application by IT Professionals, or externally as factors arising from their application in different sectors.

Use this book as a practical resource, an informative and educational source of material in developing expertise, but also as an invaluable toolkit to support practical application of ethical thinking.

Declan Brady
21 December 2021

Reference

Gotterbarn, D. (2016). Codes of ethics—the conscience of a profession: Connecting technology and society. *ACM Inroads, 7*(4), 33–35.

Declan Brady is a chartered IT professional with extensive experience working with global leaders in the provision of customised IT solutions. He is recognised for specialist expertise in Systems and Enterprise Architecture, Data Protection and Risk Management. He is President of the Irish Computer Society and a member of the Board of Directors of CEPIS—the Council of European Professional Informatics Societies, the representative body of national informatics associations across greater Europe.

Chapter 1

Setting the Scene

This is the second book in the trilogy, *Ethical Digital Technology*. The first book, *The Evolving Landscape of Ethical Digital Technology* provided an academic analysis of the subject. In the foreword to the first book, Professor Chuck Huff wrote,

> Prof. Rogerson's background in practical software development, however, has led him to a peculiar kind of interdisciplinarity. He uses ideas from all the disciplines above [anthropology, computer science, economics, mathematics, philosophy, psychology and sociology] but applies them in the practical considerations of designing and implementing real computing artifacts. . . . Simon is fixedly focused on the practical, using abstract knowledge as it is helpful.

This is the driver for the second book, *Ethical Digital Technology in Practice*, which focuses on a practical perspective. It is hoped that the reader will be drawn in to explore the accounts of the ethical and societal challenges faced by practitioners of digital technology. The future third book in the trilogy will make the subject accessible to the wider audience of both adults and children among the general public.

1.1 The Beginning of My Digital Technology Odyssey

It's November 1966. One cold, dark evening, a 15-year-old boy and his father set off to the local boys' grammar school which the boy attended. It is the school's careers evening for the fifth year. Nervous and shy, the boy walks from stall to stall, not so much disinterested but unsure what to look for and what to ask. His father encourages him to engage. After a while, the boy lingers at a stall where a man has pictures of machines which look as if they are from a science fiction film. He asks the man what they are, to which the man replies "Computers." The man shows him what computer programmes look like, how they are input using paper tape and how the results are printed on continuous sheets of paper from a line printer. He explains what computers are being used for and how exciting careers beckon. The boy, who likes making things, science fiction, physics and mathematical logic, is hooked!

DOI: 10.1201/9781003309079-1

It's May 1967. The boy is sitting in the examination hall waiting to start his English Language "O" level examination. A major part of the examination is essay writing. For years he has practised writing stories, descriptions, dialogues, and accounts, all of which have been passable but never had the task fired him. Scanning the title choices, he sees "Write a report of what the future might hold." He has never written an essay in report style and his teacher had often warned about this type of essay as it was littered with pitfalls. A sharp intake of breath and he starts writing about how the world will change with the use of computers. He finishes his essay just as the invigilator calls time.

It's August 1967. The boy opens the envelope and reads his "O" level results. English Language top grade! His essay had been successful. From that moment he knows computers are for him.

It's April 2021. I am in the process of writing this book about how the world has changed with the use of computers. It is a practical account of the computer revolution from 1995 to the current day when the world is increasingly dependent on digital technology. It explores some of the social and ethical issues that are part of this revolution. This is not a book about deep philosophical and technical concepts. Nor does it claim to be comprehensive. It is a personal account of technological change and its effects on people. It is written by the boy who was smitten by computer technology at the age of 15, became a computer professional and then spent many years trying to show young people how to develop and use digital technology in a good way. It is a book based upon my engagement with practitioners, academics, and students during the period as well as my continued fascination with this fantastic technology. This is a book about the real world, of what has happened and what might happen as digital technology continues to pervade.

In 1995 I launched a new column called ETHIcol in the *IDPM Journal*. This journal was the house journal for the Institute of Data Processing Management (IDPM) which was subsequently renamed the Institute for the Management of Information Systems (IMIS); consequently, the house journal became the *IMIS Journal* in 1997. The aim of the column was to raise issues of ethics and social responsibility in the application of ICT in a manner which was accessible to practitioners. The column ran until 2012 when IMIS was amalgamated with BCS, The Chartered Institute for IT, and the *IMIS Journal* ceased publication.

1.2 The First ETHIcol [1995]*

*This section was first published as: Rogerson, S. (1995) ETHIcol—computer ethics. IDPM Journal, *Vol 5 No 3, p. 26. Copyright © Simon Rogerson.*

This is the first edition of ETHIcol, a regular column aimed at raising issues of ethics and social responsibility in the application of computers and associated technologies. Every day, new systems are implemented in organisations in the search for efficiency gains and improved effectiveness in an effort to realise some corporate objective. Whilst these systems may be deemed successful in this context, far too often success is achieved at the price of, for example, breakdown of social groupings in the workplace, the de-skilling of jobs, over-reliance and intrusion of technology and the unacceptable side effects of system implementation. Technology has evolved from a coercive technology to a seductive technology resulting in many of these issues being overlooked. This is unacceptable and such issues must be addressed at the right time and at the appropriate organisational level.

A recent visit to a company employing service engineers throughout the country illustrates what can go wrong if implications of system implementation are not carefully and fully investigated. The company had been suffering significant thefts of its service vehicles when parked-up at night. The attraction was not the expensive, though specialised, service equipment in the vehicles but the engines of the vehicles themselves which apparently had a high resale value. The company decided to attach electronic tags to the vehicles enabling vehicle movement to be monitored from a central office. At night it was possible to place an electronic fence around the vehicles. Should an attempt be made to move the vehicle beyond the fence, an alarm was triggered at the central office and the police alerted. The system proved highly successful and thefts reduced dramatically. The management of the company then realised that this system could be used to monitor indirectly the movements of the service engineers throughout the working day, providing information about abnormal activity instantaneously and without the knowledge of the engineers. The IT manager was briefed to embark upon this spin-off system. Therein lies the problem—the legitimate use of technology giving rise to the opportunity of questionable unethical action by the company which would affect every service engineer and ultimately anyone who used a company vehicle. The IT manager was placed in a very difficult position because of the conflict in professional responsibility to the company on the one hand and to the employees as members of society on the other.

The recent Green Paper on ID cards raises many important issues particularly because of the apparent preference for a compulsory multi-faceted ID card. At a recent conference, the UK Home Secretary, Michael Howard, stated that a "multi-faceted card was a desirable ultimate objective." Very powerful technology exists enabling sophisticated ID cards to be introduced. The fundamental issue is whether they should be introduced. Such ID cards would form probably the largest distributed database system in this country with the ability to process data locally, using, for example, smart card technology, as well as transferring data to and from a number of central processing points. Much of this data is likely to be sensitive data concerning individuals. Questions of data ownership, data integrity, the ability to update and maintain data, and data access by the data subject all need to be addressed.

No technology is secure. Witness the frenetic battle by the credit card companies to stay ahead of criminals who possess the wherewithal to breach, with apparent ease, the latest security measures introduced on the credit cards. There is a tendency for many people to accept computer generated information without question. Forged computer-generated ID cards provide the criminal with a way to legitimise their identity when attempting to gain access to homes on the pretext of being an authorised tradesman. Inevitably, we will come to rely on ID cards as the ultimate proof of identity. This digital icon will be all important, without it the citizens of society will find it increasingly difficult to live and work—they will become the non-citizens whose privileges and opportunities will be forfeited.

Such issues are not isolated. The technology revolution, based upon a logically malleable computer core, presents more and more opportunities to utilise technology in the quest to achieve more and more ambitious goals. Computer professionals must accept their responsibilities in this revolution and must persuade organisations to apply the computer technologies in an ethically sensitive way—a way which is acceptable to the citizens of society.

1.3 Ethics Man [1999]*

*This section was first published as: Middleton, C. (1999) Ethics Man. Business and Technology, January, pp. 22–27. An edited version, on which this section is based, was published in: This Quarter. De Montfort University, Issue 6 Summer, 1999, pp. 20–23. Copyright © Chris Middleton. Reprinted by permission.

In this article, Chris Middleton talks to the UK's first professor in Computer Ethics about technology versus humanity.

We live in an age where business decisions are made by the discreet whir of a CD-ROM drive and where human beings are fired at the flick of a wrist on a mouse mat. Technology, perhaps more egalitarian and democratic than ever before, also carries with it unprecedented means for the (literally) systematic abuse of employee rights.

Just the stroke of a key can alert employers to our diligence; genetic fingerprinting contains information about us so intimate that we ourselves are unaware of it, with the desire of supermarkets to stack the shelves of their data warehouses means that we really are what we eat, wear and sleep in—thanks to our loyalty cards.

But who cares? Simon Rogerson of Leicester's De Montfort University does. With the aspect of Rodin's Thinker, he looks down on the technology-led workings of British business and muses on their ethical implications.

Rogerson, who describes himself as "positive, entrepreneurial, socially aware, inclusive and humorous," is director of the Centre for Computing and Social Responsibility, which makes him the UK's first professor in computer ethics. Now 47, he is a graduate of computational science from the University of Dundee.

"People might be surprised that a professor in computer ethics has a non-philosophical background and began his working life writing Fortran and Assembler programs," he says. Whether they are surprised or not, Rogerson clearly has a credible foundation for what cynics might see as an airy discipline.

"As computer technology advanced, people started to be aware of the pitfalls that threatened to undermine the benefits of this powerful resource," explains Rogerson. But how can a classical model be applied to a 20th-century phenomenon?

Rogerson is studying an entirely new set of problems, he explains, such as fraud and computer-generated human disasters, and the ways in which they present new versions of standard ethical dilemmas. In academic terms, this means embracing concepts, theories, and procedures from philosophy, sociology, law, and psychology as well as computer science and information systems. "The overall goal is to integrate computing technology and human values," he says, "in such a way that technology advances and protects human values, rather than damages them."

And technology advances at an astonishing rate, from back room to living room via the desktop and the TV. Five years ago, he points out, if someone had said they could shake hands with somebody 1,000 miles away, people would have laughed. Today the technology exists to do exactly that, but legislation has yet to come to terms with how people can interact from opposite sides of the world.

Does this suggest there is something about technology that tempts businesses away from shared ethical foundations? Rogerson disagrees. "It's not that ethics are being put aside when it comes to ICT, it's just that they are only now being recognised by businesses as an issue.

"Computing crosses cultural, religious, political and economic boundaries, and, as such, challenges our social norms. But there are several core values that are common, such as knowledge, freedom and impartiality, on which a universal code for ICT can be founded."

"There are so-called 'policy vacuums' created by technology which lead to ethical dilemmas," admits Rogerson. "Some may arise from new twists on old problems." Such as?

Privacy is a good example. Ever since civilisation began, the right to privacy has been a philosophical issue. In a recent speech at the Business Link annual conference, Peter Mandelson said, 'The key to our future competitive success lies more and more in the exploitation of knowledge for commercial, profitable ends.' The concern is how this will be translated into organisational strategy and action.

Another, surely is that capitalism relies on competition to supply people's needs. But market forces are reductive, and if the market merely regards us, say as ABC1s who shop at Sainsbury's, then the commercial exploitation of that knowledge could mean nothing less than an in-depth exploitation of people. It's as simple as ABC.

> Knowledge resides with people, and I can imagine an organisational world where people are cherished, encouraged and justly rewarded so they remain content and loyal. But I can also imagine one where people are sucked dry of their knowledge so it can be retained in an ICT system while they are discarded as another spent resource.

As businesses are discovering, fraud detection lends itself to data-matching systems that need little or no human intervention, and commercial pressures mean that the use of such systems will grow. (As is the case with most technologies; witness biotechnologists" pressure on the Government to decide its ethical stance on human cloning so industry can plough ahead and make money.)

Today, many financial systems are wholly automated, and tens of thousands of people are now being laid off in the City as a consequence. What then, are the ethical repercussions of business systems that "mine" for knowledge or "predict" patterns of human behaviour from a database of existing information?

> There is a difference between the detection methods used in the past and today's data matching. Traditional investigation is triggered by evidence of wrong-doing by an individual, such as tax evasion or bogus benefit claims. Data matching, however, isn't targeted at individuals, but at entire categories of people. It isn't initiated by suspicion about an individual, but because the profile of a particular group is of interest. The data-matching process reverses the assumption of innocence.
>
> Data held legitimately in the public domain should be allowed to be traded. But bringing together private, personal data using, for example, an automated inference model (such as software that predicts you might commit fraud) may create false knowledge which, when traded, results in harm to the individual. That mustn't be permitted.

The professor claims that, when it comes to computing and business, morals and ethics are interchangeable. Morals originate from the group in which a person matures, and business ethics, by implication, from the group in which your business matures. This is why, he says, organisations must promote ethical practices to compensate for the "policy vacuums" he talks about. In other words, to create a shared ethical culture among their peers.

In the business world, of course, employers often argue that electronic monitoring deters fraud, industrial espionage and other illegal activities, but this does not, says Rogerson, give them a universal right to monitor their employees. The civil liberties of innocent people should not be suppressed because a few rogue employees might abuse them.

In the light of all this, it's easy to see why De Montfort set up the Centre for Computing and Social Responsibility in 1995. It has generated enormous interest: the centre's website notched up nearly 20,000 hits on the day we spoke in early December, and recorded 275,000 visits for the whole of November. But what persuaded the university that Rogerson was the right man for the chair?

"Computer ethics must be practically relevant," he says. "So, my background in practical computing makes me credible in the eyes of the IT user."

With the groves of academe subject to ever-greater market forces since the Thatcher era, is the department itself ethically compromised? Who is sponsoring Rogerson's research?

> Our work involves many of the leading organisations in the UK, including Royal Mail, BT, IBM, Transco, the Institute of Business Ethics, the Institute for the Management of Information Systems, the British Computer Society and the office of the Information Commissioner. Any organisation using ICT should be interested in our work and, more importantly, should be committed to the concept of computer ethics. I would like to see a cross-section of organisations working with us, representing various sectors and coming from different cultures and countries.

Many of these organisations, he says, participate in workshops and research agendas that work towards social responsibility in the electronic age, with the idea of creating something of practical benefit to the world at large, rather than mere theorising. To this end, the Economic and Social Research Council have funded a research project with Rogerson's department. "IS IT Ethical? The 1998 Survey of Professional Practice" published in February 1999 looks at the ethical perceptions of information systems managers. It reveals an IT community split along lines in response to ethical questions.

One area of particular interest to Rogerson is the subject of our very identities in the digital age. We have become, he says, composite beings with electronic personas, which present new problems of identification.

> We run the risk of creating a two-tier society; the "citizens," and a second group, the "underclass," who don't have access to the digital society or any of its services. As I see it, there are now three elements to us as individuals: the physical, the philosophical and the digital.

So, what advice does the professor have for the UK's IT directors and strategists? "The impact of IT is usually judged in terms of whether planned gains in efficiency and effectiveness are realised, "he says.

> But that isn't everything. My advice would be to consider who is affected by your work; examine if others are being treated with respect; consider how the general public would view your decisions; and analyse how the least empowered will be affected by your decisions and consider if your actions are worthy of a model IT professional. Good ethics mean good business.

1.4 Making a Difference [2009]*

*This section was first published as: Himma, K. (2009) Simon Rogerson: making a difference 2005. ACM SIGCAS Computers and Society, Vol 39 No 2, pp. 22–23. Copyright © the Association for

Computing Machinery Reprinted by permission. Simon Rogerson was the author of most of this article, having written all the Answers.

Ken Himma [Qn] interviews Simon Rogerson [A], the recipient of the ACM SIGCAS Making a Difference Award in 2005.

[Q1] Looking back, what was it that first sparked your interest in computer/information ethics?

[A] Having spent many years in industry working on systems development in a variety of programming, analysis and project management roles I had my usual set of battle scars and medals borne out of trying to get systems to work and produce useful output for client departments. When I changed careers I was first employed as a senior lecturer and my role was to instil a sense of IT reality in final year computer science undergraduates and master's level postgraduates. The problem of addressing system success and failure was high on my agenda as I had experienced little of the former and a lot of the latter in my time in industry, which was a typical profile for any IT professional at the time. By the early 1990s, I became convinced that the problem was that IT people were looking at system development with a very narrow perspective. What was needed was a much broader scope. It was then I discovered some of the work being done in computer ethics. Like many of us working in this field it was Deborah Johnson's writings that I first discovered and started to use. I found it fascinating that here was I very much a practical IT professional looking outwards for solutions and that there were several academics, primarily from philosophy, looking inwards to try to explain the phenomena they observed. I was hooked!

[Q2] In your opinion, what is/are the most pressing issues in our field today? Why?

[A] Converging technologies continue to change the world we live and work in. The impact across the world grows but still only around 20% of the global population use information and communication technology (ICT). The disparity of opportunity to benefit from ICT between poor and rich, under-developed and developed, and rural and urban remains and in some cases has increased. Associated with this are the power structures that have been created with advancing ICT. Those who own and/or control ICT infrastructure, media outlets and application systems wield much power in the modern world. It is they who decide whether ICT is good for us, what are the ICT priorities and when it is time to cease non-ICT products and services. For these reasons we need to assess new technological advances, for example cloud computing, implants and non-human agents, as well as existing ICT usage. We need to ensure those working in ICT understand and accept all their professional obligations and responsibilities and provide them with the instruments to do so. Our work in the field must be accessible to policy makers, industry, educators and the public for it is too important to remain in the dusty corridors of academia.

[Q3] Where do you see the field in the future?

[A] We must never forget that ICT is a practical domain and whilst it is important to address issues with rigor we must never see it as simply a vehicle for exercising and exhibiting our intellectual cleverness. I remain convinced that whilst exploration of the conceptual foundations is essential, this must always be done with the understanding that it will help to ensure that the development and use of advancing information and communication technologies benefits us all and does no harm. There continue to be differing opinions about the nature of the field on a continuum from a philosophical-only position to a wide interdisciplinary position. For me the issue is quite simple—we live in an interdisciplinary world, we use resources in an interdisciplinary way, and we face issues, challenges and problems that require us to draw upon our interdisciplinary skills and experiences. ICT is just part of our world and as such demands to be treated in an interdisciplinary way. So, the field must be interdisciplinary—it is this that has coloured the way in which we have operated the ETHICOMP conference series since 1995. Through ETHICOMP new interdisciplinary partnerships have evolved. In Europe, research funding has recognised the need for interdisciplinarity and practicality. More emphasis is being placed on funding ethics-oriented ICT projects but only when the approach is interdisciplinary and applied.

[Q4] What advice or words of encouragement would you give to up and coming scholars?

[A] For the foreseeable future ICT will remain at the forefront of human advances. Its pervasiveness is breathtaking. It is an amazingly stimulating and challenging area to work. It is particularly so for those of us focusing on the ethical and social impact of ICT. It requires us to be open-minded in our pursuit of answers drawing upon any existing relevant knowledge regardless of source. One of the greatest joys for me has been the engagement with scholars from many different disciplines, countries and cultures who have come together in a supportive way with one simple goal to make ICT better. We need young scholars to take up the mantle of pioneers in the field and take us forward. Whichever discipline you come from there is a place in this field for you if you are willing to adapt and to work in partnership with those from other disciplines. Our community encourages young scholars to engage and lead—it is really true that if you are good enough you are old enough.

[Q5] Is there anything else you would like to share with our readers?

[A] I have just come back from delivering the Social Impact of Computing Summer School for master's computing students at Gdansk University of Technology in Poland. All these students worked fulltime in the industry. This is the first time they had been exposed to the broader issues surrounding ICT. At the end of the course, I asked them what they had learnt and would take back to their work. Many of them told me that they had never thought about how what they did in ICT might harm people or the environment but from now on they would. For me this is why our field is important, this is why we must engage in the delivery of education, this is why we must continue to lobby and this is why we need new scholars to join the field.

1.5 ENIGMA [2012]*

This section was first published as: Rogerson, S. (2012) ENIGMA: Reflections on Bletchley Park. National Trust Peak District Centre Newsletter, June, p. 3. Copyright © Simon Rogerson.
    ENIGMA start of message: NNGQA LKYAN LDLJL XADH[1]
    A group of intrepid would-be code breakers set off from the Peak District on Saturday morning 19 May 2012 bound for Bletchley Park. Led by our intrepid leader Marie Ware, whom we all thank for such a fantastic day, we arrived late morning. After a brief introduction we dined on shepherd's pie and mixed veg which got us into a wartime frame of mind. There we were in the drawing room of the mansion where intellectual heavyweights such as Alan Turing, Dilly Knox,

---

[1] This is an example Enigma code generated using my smart phone app.

John Jeffreys, and Max Newman must have once sat to relax and talk as they worked on cracking Nazi codes and cyphers. The first code breakers, masquerading as "Captain Ridley's Shooting Party" to disguise their true identity, arrived in August 1939. Eventually there would be 10,000 people at Bletchley involved in this vital work. Two thousand messages were cracked each day at the height of activity.

For me as a computer scientist, it was a special day to visit, in essence, the birthplace of modern computer technology. Our enigmatic father and son guide duo took us on a fascinating journey of understanding. They explained what the Enigma machine was, its significance in WW2 and how the generated coded messages were cracked. The Enigma machine was a mechanical implementation of a poly alphabetical cypher transformation which basically scrambles words into strings of five letters which then cannot be deciphered without knowing the complex key settings. The key meant that there were 158 million, million, million Enigma variations making it possible to create very secure coded messages.

It was the search for this "shared secret" of the key which I think was the most important work of the code breakers. This was eventually done using the Bombe machine designed by Turing, which comprised 36 Enigma machines working in tandem, to search through all the permutations in order to try to identify the key settings. It typically took 30 minutes to find the settings and then the messages could be deciphered. The successful combination of practical ingenuity and intellectual brilliance pervades Bletchley. Turing's design was turned into reality by using off the shelf components so that many Bombes could be quickly built.

The other highlight of our trip was Colossus, the world's first programmable electronic computer. It was this that helped to crack the Nazi's strategic cyphers. Designed by Tommy Flowers, each of the ten that were built could crack codes in about six hours. This enabled the allies to gain access to fine detail of the German defences. It is recognised today that the code breakers work probably shortened WW2 by at least two years.

To see the rebuilt Bombe and the rebuilt Colossus operating was amazing. Back home I downloaded on to my smart phone my own enigma machine so I can send coded messages one of which I have shared with you here. Without Bletchley I would not have a smart phone and I would not have my own enigma. This is an "app" way to finish my account!

ENIGMA end of message: THIS WAS A GREAT DAY OUT[2]

1.6 Outline of the Book

This book comprises seven chapters; Setting the Scene (this chapter), Data and Information (2), Systems and Applications (3), Practice and Method (4), Regulation and Policy (5), Practical Considerations (6), and Synthesis (7). Articles have been assigned to specific chapters using the primary focus of each article. Data and Information (2) and Systems and Applications (3) focus on the scope of digital technology, whereas Practice and Method (4), Regulation and Policy (5), and Practical Considerations (6) focus on the implementation of digital technology. However, many articles do cover several perspectives and, hence, have relevance in other chapters. Together the seven chapters describe an evolving landscape of digital technology, society, organisations and people.

ETHIcol is the foundation of this book and as such most of the columns are included. Each column is approximately 1,000 words in length. Additional materials, mainly between 2012 and

---

[2] This is the decoded Enigma code.

2021, are from occasional articles in other practitioner-oriented magazines as well as blogs and opinion pieces posted on websites. Most of the articles are written in a journalistic style presented in short paragraphs, but there are some which are written in a more formal academic style. The section title includes the date of the original publication (shown between "[]"). Where a section has been published before full reference details are included as a footnote. Wherever possible, published papers retain their original form to provide authenticity in terms of terminology, subject matter, perception and opinion at the time of publication. The sections in each chapter are presented in chronological order. As such, these papers become a reflection of the evolving technological revolution and its impact on the world and across society. The book concludes with a series of overviews which reflect practitioner opinion at different points along the timeline covered by the book. Finally in the Afterword, a way forward is suggested which promotes digital technology that is fit for purpose, accessible and acceptable.

Chapter 2

Data and Information

Digital technology is primarily concerned with capturing and storing data which is then transformed, through organising, structuring, and presenting, into accessible information within given contexts. Information leads to knowledge which enables judgements that, in today's world, can be taken by humans and machines. The pervasiveness of digital technology has led to the mushrooming of data generation. Statista (2021) reports that 74 zettabytes (1 zettabyte is 1 trillion gigabytes) of data will be created globally in 2021 compared with 59 zettabytes in 2020 and 41 zettabytes in 2019. It is projected that this will rise to more than 180 zettabytes by 2025. Such incredible data generation inevitably leads to not only information, in its many forms, but also misinformation (false, inaccurate, or misleading) and disinformation (deliberately deceptive). It is unsurprising that systems, organisations and individuals suffer from information overload and information pollution. The provenance and integrity of data and information become increasingly important as data generation escalates. These issues are explored in this chapter with accounts of how the data-to-information transformation process, and accompanying challenges, have changed over time.

2.1 How to Create Waffle! [2000]*

This section was first published as: Rogerson, S. (2000) ETHIcol—How to create waffle. IMIS Journal, Vol 10, No 4, pp. 29–30. Copyright © Simon Rogerson.

Ingredients

- one internet search engine of your choice
- two or three keywords
- one hour of your time

Method

- open your internet access
- activate your World Wide Web browser
- go to the internet search engine of your choice

DOI: 10.1201/9781003309079-2

- send two key or three keywords about the information you are requiring
- wait for the list of websites to be presented (the list is likely to contain many thousands of websites)
- for each of the first ten websites go to the site and collect the information about your keywords
- produce a summary of what you now know about these keywords

Outcome

- waffle for use the next time you are stuck for something to say!

Is this a familiar recipe? How many times have you been searching for those key pieces of information and been presented with a bewildering list of websites which appear to have little if any connection with the key words you had entered? It is a recipe for disaster and unfortunately, it is all too common. It is a symptom of the Misinformation Society and in some ways we all bear responsibility for we have become information junkies who feed on the byte-size trivia that many of the so-called internet information engines provide. Indeed, many of the traditional information sources have become backwaters left to stagnate through lack of investment.

Does this lack of appropriate information really matter? It might be annoying and it might make us a little less effective and efficient but is there more to this modern way to be?

In a recent article (*The Independent Monday Review* 19 June 2000), Suelette Dreyfeus discussed the concerns about computer-based information. She reported that leading neuroscientist Professor Susan Greenfield believes that computer-based information could result in "loss of imagination, the inability to maintain a long attention span, [and] the tendency to confuse facts with knowledge." Greenfield's concern is that those without life experience will be unable to cope with the bombardment of information because they will not have a cohesive framework in which to place so much information. All these facts do not produce wisdom. This is done by relating facts and then reflecting and inferring. Greenfield argues that the sanitised Information Society with its information overload may well be changing the way in which we think and may well be reducing our ability to assimilate information, become knowledgeable and undertake judgement.

It is clear that such concerns will have a growing impact on information systems professionals. The information systems professional has, for many years, been charged with providing systems that deliver information for a continuum of users from members of the general public to key individuals within organisations. These professionals have been trained to produce systems which deliver information that is timely, relevant, accessible, and accurate. As many reports of system failures demonstrate, such information is not always delivered and, with the advent of the internet and intranets, misinformation appears to be more and more likely given the typical scenario outlined at the beginning of this article.

We have a professional responsibility to guard against this. Indeed, the draft code of ethics for IMIS (published in a previous edition of the IMIS Journal) states that "Every Fellow and Member of the Institute (including both Professional and Affiliate Membership grades) shall employ his or her intelligence, skills, power and position to ensure that the contribution made by the profession to society is both beneficial and respected." It further states that the IS professional should

> strive to ensure that professional activities for which I have responsibility, or over which I have influence, will not be a cause of avoidable harm to any section of the wider community, present or future, . . . [and] . . . use my knowledge, understanding and position

to oppose false claims made by others regarding the capabilities, potential or safety of any aspect of Information Systems.

It is therefore clear that all information systems professionals must consider carefully the provision of so-called information, ensuring that it not only satisfies the immediate need but that also the manner in which this is achieved is not to the long-term detriment of those using such systems. Perhaps it is time that information systems professionals join with the information science professionals, the traditional custodians of information sources, to produce a rich and valuable information resource that will educate and inspire all, helping them to develop as individuals.

2.2 Information for All [2001]*

This section was first published as: Rogerson, S. (2001) ETHIcol—Information for all. IMIS Journal, Vol 11 No 3, 2001, pp. 25–26. Copyright © Simon Rogerson.

This edition of ETHIcol draws from participating in the Information Society and Intelligent Information Technologies in the 21st Century conference held in Moscow in April 2001. In his address to the conference, Ivan Sergeev, Deputy Minister of Foreign Affairs of the Russian Federation, explained that the Information Society needed to recognise cultural diversity and a multi-polar world structure. If we are to achieve an inclusive Information Society then this must be taken into account.

The key issue is one of enabling citizens. The recent report from The British Council, entitled "Developments in electronic governance," suggests an evolution from passive information giving to an active citizen approach. Conditions for participation are categorised as:

■ enabling participation in the Information Society
■ creating the infrastructure for the Information Society
■ fostering a sense of citizenship and cultural identity using ICT

In a previous edition of ETHIcol a series of six questions was posed about the reasonableness of computing applications. These can be used to consider the issues surrounding building an inclusive Information Society based upon the three conditions above. So does the Information Society:

■ promote social and economic justice?—Clearly the emergence of electronic governance has great potential to promote such justice. The proposal by Citizens Online to create a Civic Commons in Cyberspace is a good example of realising this potential. However, we must be aware that such advances will change the relationship between citizens and government. The ramifications must be considered very carefully.
■ restore reciprocity rather than consolidate power in the hands of the minority?—The first point leads on to the issue of reciprocity. Media moguls are already wielding great power in the Information Society. Unhampered access to the on-line world can become very difficult in this situation. It is unclear how the Information Society will unfold now there are a small number of big players controlling strategic elements of content and infrastructure.
■ benefit the many rather than the few?—Currently the vast majority of the facilities are available in English using a standard interface and presented with a Western culture bias. It requires a high level of literacy and numeracy to navigate to a chosen facility. Much more must be done to ensure access to people who do not have English as their mother tongue or are of lower intellect or have a physical or mental restriction or are from a different culture.

- put people first rather than the technology?—The tools of the Information Society are still primitive. The day when the technology is transparent will be the day people have been put first. Gone will be the need for identity numbers, digital signals and the like. Gone will be the need to be computer literate and dexterous.
- limit economic gain because of potential social and environmental cost?—The much heralded efficiency gains through computerisation become questionable as more socially sensitive applications are implemented. The abolition of data redundancy may appear to be key in efficient computerisation but may be unacceptable if having only one occurrence of a data item causes socially and ethically unacceptable data relationships within the system in question.
- favour the reversible over the irreversible to ensure impact of rogue ICT applications can be reversed?—Systems of the Information Society require massive investment. This means traditional systems are likely to be left in poor state of repair or may be phased out completely. There is no going back and there is no choice in such situations. This may prove disastrous for society. Perhaps the Information Society should be developed as having complementary systems—some computerised and others non-computerised.

Clearly there is much to be done in realising the inclusive Information Society. What are you going to do by way of contributing to universal access so all of society can flourish in this new age?

2.3 Information Provenance [2005]*

This section was first published as: Rogerson, S. (2005) ETHIcol—Information Provenance. IMIS Journal, Vol 15 No 1, 2005, pp. 33–34. Copyright © Simon Rogerson.

In January 2005 the UK public acquired five important new rights to information held by public authorities.

- **The Freedom of Information Act 2000** comes into force, after a four-year delay to give authorities time to prepare. The Act applies to central government bodies and to English, Welsh and Northern Ireland public authorities. It also applies to the House of Commons, the House of Lords and to the Welsh and Northern Ireland assemblies.
- **The Freedom of Information (Scotland) Act 2002** applies to the Scottish Executive, the Scottish Parliament and Scottish public authorities.
- **The Environmental Information Regulations 2004** provide a separate right of access to environmental information held by UK public authorities. Some private bodies, including utilities and contractors providing environmental services on behalf of authorities, are also covered. The regulations implement an EU directive.
- **The Environmental Information (Scotland) Regulations 2004** provide a similar right of access to environmental information held by Scottish public authorities and certain private bodies.
- **Amendments to the Data Protection Act 1998** strengthen people's rights to see personal information about themselves held by public authorities throughout the UK, including Scotland. The Act already allows people to see computerised personal data about themselves and medical, social work, housing and school records. The amendments significantly improve the right to see other paper records.

(source: The Campaign for Freedom of Information, Press release: 31 December 2004, www.cfoi.org.uk/foi311204pr.html accessed January 2005)

These rights give a new impetus to raising awareness of the need for information integrity and the implications of the lack of it. Information integrity is about accuracy, consistency, and reliability of information content and information systems. Madhavan Nayar (26 July 2004) explains that,

> Digital information is becoming as pervasive and essential as air, water, electricity and canned food. Increasingly, we rely on such information for our livelihood, lifestyle and even life itself. Ironically, however, information has not been the focus of interest thus far in this information age.

If such information is questionable then decisions and actions which are based upon it could be flawed and unsafe. How many situations like this will come to light when people gain access to information using their new rights? How many situations like this remain hidden in the organisations not covered by the new legislation?

The expectation that digital information is dependable and trustworthy is reasonable. But how can dependability and trustworthiness be demonstrated? Trustworthiness is an intrinsic reality. Its perception, particularly in the beginning, depends critically on the perception of certain extrinsic forms (signs, labels, messages, etc.) that are understood to represent the presence of underlying trustworthiness. In the case of digital information these extrinsic forms should represent the *Information Provenance*. Information provenance fixes the origin and network of ownership thus providing a measure of integrity, authenticity, and trustworthiness. It provides an audit trail showing where information originated, where it has been and how it has been altered. In this way people would be able to consider how much credence they would give to a piece of information before acting upon it.

Consider this example. In the course of its enquiries a police authority collects information about an individual. This information is held within the police authority's information systems. Such information is allowed to be shared with a number of other authorised agencies across a secure network. Access is instigated by the agencies so no track is kept of where the information has been shared. Once this happens the copies of this information become legally owned by the recipient agencies. Agencies update this information for their own purposes and based upon their own intelligence. These new versions of the information are passed onto other authorised agencies. The police authority then updates the information about the individual based on new evidence. Agencies are not aware of this and continue to use their own version of the information.

In this situation there exist multiple copies of the information across a complex network of agencies. Copies are not the same and there is no mechanism in place to ensure that they are the same. Clearly the integrity of the information is questionable but those receiving it are likely to be unaware of this. Decisions may be made are based on this untrustworthy information, that have detrimental effects on the individual. If the information had been accompanied by the information provenance then decision makers would be able to see how the information had changed and therefore consider how safe it was. Also, provenance would provide a method to track back to the provenance of original information held in the information systems of the police authority to check whether the original information had altered since it was first accessed.

From this example which is based upon a real situation it can be seen the information provenance is a powerful instrument in improving information integrity and trustworthiness. According to Fox and Huang (2003) there are four levels associated with information provenance

- ■ **Level 1 Static** This focuses on provenance of static and certain information.
- ■ **Level 2 Dynamic** This considers how the validity of information may change over time.

- **Level 3 Uncertain** This considers information whose validity is inherently uncertain.
- **Level 4 Judgment-based** This focuses on social processes necessary to support provenance.

If such concepts of provenance were incorporated into the practices and processes around information then the management and governance of information would be much improved. Just think—if every piece of information was accompanied by its provenance what a different Information Society we would live in.

2.4 Information Integrity in the Information Age [2011]*

This section was first published as: Rogerson, S. (2011) Information Integrity in the Information Age. In: Haftor, D. & Mirijamdotter, A. (eds.) Information and Communication Technologies, Society and Human Beings: Theory and Framework (Honoring Professor Gunilla Bradley). IGI Global, pp. 329–335. Copyright © Simon Rogerson.

2.4.1 Introduction

Converging technologies have changed the way we should look at information. They have raised a set of fresh issues which need to be fully explored and addressed if we are going to realise the full potential of the Information Age. The growing dependency on information and communication technologies (ICT) to publish, consume, and manipulate information has impacted upon economic and cultural life. The Information Age has spawned a new society—the Information Society. The Information Society crosses traditional boundaries and as such comprises individuals from many different cultures. This cultural variability means that the expectations of individual cybercitizens can differ considerably.

Everyone belongs to one or more cultural communities and so, as Hongladarom and Ess (2007) point out, it is impossible to consider the Information Society from a culturally neutral perspective. For example, Brey (2006) describes a positive view of the internet from a libertarian ideology which is in marked contrast with the negative view he describes from the perspective of Orthodox Judaism. This cultural variability has been explored by Nance and Strohmaier (1994). They suggest there are two important dimensions to consider regarding cultural variability. The first dimension is the continuum from individualism to collectivism. Individualism emphasises self-interest and promotes the self-realisation of talent and potential. Its demands are universal. Collectivism emphasises pursuit of common interests and belonging to a set of hierarchical groups where, for example, the family group might be placed above the job group. The demands on group members are different to those on non-group members. The second dimension concerns cultural differences in communication referred to as low context communication and high context communication. In the former, the majority of the information resides in the message itself whilst in the latter, the communication is implicit. Nance and Strohmaier (1994) suggest that the US utilises low context communication whilst Japan uses high context. So even with a shift towards cultural homogenisation through ICT usage, the variability that remains makes it very challenging to provide information or conduct a debate in a way that is acceptable to all (Fairweather & Rogerson, 2003). It involves establishing a set of common behavioural standards whilst ensuring that there is no dominant participant. The current internet seems a long distance from this position. Indeed, given such cultural variability, it is clear that there are great difficulties in providing information in a form that is acceptable to all. This is certainly one of the great challenges of the Information Age where we all create, communicate, and consume information.

2.4.2 The Nature of Information

But what is information? Eaton and Bawden (1991) suggest that information exhibits five intangible characteristics which differentiate it from other types of resource. The value of information is difficult to quantify, and its value is relative in terms of both time and information users. Information has a multiplicative quality in that it is not lost or decreased if it is consumed and indeed using information often causes it to increase in value and size. Information is a dynamic force within the system it resides which causes that system to flex. There is no predictable life cycle of information, indeed once-dormant information can become current and valuable in changed circumstances. Finally, information manifests itself in different forms relating to particular situations. Meyer (2005) extends this list to include other characteristics. Information has the attribute of alleviating uncertainty. Information is always an essential intrinsic component of technology. Information is a catalyst to enhance economic growth. Information extends the knowledge base. As Macgregor (2005) states, "Ultimately information behaves in a unique manner when compared to other resources because it essentially represents the genesis of human thought, and is heterogeneous and intrinsically intangible."

Technological maturity has reached a point where information of nearly every form is available at the touch of a button, the click of a mouse or the pointing of a cursor. Never before has it been possible to support many-to-many or many-to-one information publication and consumption. Existing new information conduits such as blogs, podcasts, and wikis offer so much. Access is now possible to all forms of information including music, moving images, literary works and art. Consider the following example of information conveyed in different forms. This has been used by the author to explore the relationship between ICT and information with students at different levels and in different countries.

On 26 April 1937 the Basque town of Guernica was subjected to the first carpet bombing of a civilian population during the Spanish Civil war. The raid was conducted by the German Luftwaffe "Condor Legion" and the Italian Fascist Aviazione Legionaria. The attack destroyed the majority of Guernica and there were widespread civilian deaths. The attack is recorded in military books as an example of terror bombing. It is still a source of emotion, revulsion, and public recrimination. In 1937, Picasso painted his famous anti-war painting. This 3.5 metres by 8 metres canvas in black and white depicts the horrors of war in a way never seen before.

The events of Guernica are an invaluable information source for the education of future generations. The challenge is how to use it. Can ICT be used to assist in its educational value? Here are some suggestions.

- Picasso's Guernica is accessible in electronic form and therefore can be displayed using PowerPoint and a data projector. Against this backdrop the story of Guernica can be told and the audience given a sense of what it was like to be there by introducing the sound of bombing as the story unfolds.
- There is an evocative song by Katie Melua called *Market Day in Guernica* which tells the story of the day of the bombing. This can be used as a soundtrack to accompany a set of still photographs of the destruction including pictures of distraught children in the bombed out streets. A period of silent reflection follows for the audience to think about what they have seen.

■ There are many accounts of Guernica on the web which can be used as sources to deliver a more factual account of the events. This can be used to explore, in an objective way, the reasons the bombing took place, its impact on the civil war and how it has been used since to promote particular causes.

These three, very different approaches provide information to the audience. Two make use of converging media to capture the attention of the audience and shock them into thinking more deeply. It is information but not in a form that we perhaps recognise. The last version is a traditional way of information dissemination. ICT-based information has a history of text and words and this final approach relies upon this.

This example and its use by the author in a variety of settings illustrate how ICT can be used to create and convey information in a variety of ways. But what does the introduction of ICT do to the meaning of information? Does the reduction of a 3.5 metres by 8 metres canvas into a small projected image in a lecture theatre detract from the painting's impact? Does the adding of sound detract or add to the impact? Is the creation of a slide show with a popular song accompaniment simply an emotional trick or does it have informational value? Will the bullet-pointed objective presentation tease out the real information rather than dwell on the emotional hype and retain the audience's attention?

These questions lead to some general statements about ICT and information. ICT offers new forms of access to information. However, ICT has the potential to add value as well as subtract value to existing informational forms. The consumption of information and the resulting value derived are greatly influenced by individual perception and learning styles. Often the perceptions of the information creator are different from those of the information consumer. It is concerning that these types of issues are rarely considered when "enhancing" information provision through ICT.

It is the power and versatility of ICT which can change the nature of information and how we perceive it. Borgman (1999) suggests there are three types of information. There is information about reality in which reports disclose what is distant in space and remote in time. There is information for reality in which recipes transform reality and make it richer materially and morally. Finally, there is information as reality in which recording information through the power of technology steps forward as a rival of reality. It is the latter which challenges the traditional view of information. Borgman illustrates this third form with classical music.

> The technological information on a compact disc is so detailed and controlled that it addresses us virtually as reality. What comes from a recording of a Bach cantata on a CD is neither a report about the cantata nor a recipe (the score) for performing the cantata; it is in the common understanding of music itself.

2.4.3 *The Information Dichotomy*

There is an interesting dichotomy in the technologically-dependent Information Society. Information is the lifeblood of organisations in the information age (Rogerson & Bynum, 1995). The veins of data communications along which this blood circulates are the new utility of the Information Society. Without communicated information organisations cannot interact with individuals and other organisations along the supply chain. However, with the advent of computer technology and more significantly the convergence of this technology with other technologies such as media, the amount and type and amount of information available has exploded. Toffler (1970) predicted this information overload where individuals and organisations were swamped with so much information that it prevented decision making and actually reduced knowledge. This problem continues to grow at a seemingly accelerating rate. Indeed, Nielsen (2003) argues that we are reaching the point of saturation, "Information pollution is information overload taken to the extreme. It is where it stops being a burden and becomes an impediment to your ability to get your work done."

If humankind is going to survive this mutation of information lifeblood into information pollution a new way of thinking and an associated new way of operating have to be derived. As Evans remarks (1979),

> Computers, in other words, have not arrived on the scene for aesthetic reasons, but because they are essential to the survival of a complex society, in a way that food, clothing, housing, education and health services are essential to a slightly simpler one. The truth is that one of the main problems—perhaps the main problem—of the time is that our world suffers from information overload, and we can no longer handle it unaided.

2.4.4 Integrity and Provenance

One aid that could reverse this mutation is explicit guidance as to the status of information as it is presented. In other words, to provide a rating of the integrity of the information before it is consumed. Information integrity is about accuracy, consistency, and reliability of information content (Mandke & Nayar, 2004) and information systems. If information is questionable then decisions and actions which are based upon it could be flawed and unsafe. The expectation that information has integrity and therefore is dependable and trustworthy is reasonable. But how can dependability and trustworthiness be demonstrated? Trustworthiness is an intrinsic reality. Its perception, particularly in the beginning, depends critically on the perception of certain extrinsic forms (signs, labels, messages, etc.) that are understood to represent the presence of underlying trustworthiness (McRobb & Rogerson, 2004). It is these extrinsic messages which would provide the much-needed guide to information integrity.

If such messages were recorded over time then the information would exhibit a provenance. In general provenance defines the place of origin and is a proof of authenticity or of past ownership. Therefore, information provenance fixes the origin and network of ownership thus providing a measure of integrity, authenticity, and trustworthiness. It provides an audit trail showing where information originated, where it has been and how it has been altered. In this way people would be able to consider how much credence they would give to a piece of information before acting upon it. For any piece of information people should be able to answer, "Can this information be believed to be true? Who created it? Can its creator be trusted? What does it depend on? Can the information it depends on be believed to be true?" (Huang & Fox, 2004). According to Huang and Fox (2004) information provenance has four levels; Static, Dynamic, Uncertain and Judgement-based. The Static level focuses on provenance of static and certain information. The Dynamic level considers how the validity of information may change over time. The Uncertain level considers information whose validity is inherently uncertain. Finally, the Judgment-based level focuses on social processes necessary to support provenance. This model enables information to be categorised which in turn provides integrity guidance. Moreau et al. (2008) suggest that ICT applications must be provenance-aware so the information's provenance can be retrieved, analysed and reasoned over. This seems to be regardless at which of the four levels the information resides. They cite the example of an organ transplant management system which comprises a complex process involving medical decision making, data collection, organ analysis, and eventual surgery. If this system was provenance-aware and had embedded in it a provenance life cycle then on demand queries such as: list all doctors involved in a decision; find all blood-test results for a donation decision; or find all data that led to a decision, could be easily satisfied. This would ensure system integrity and increase trustworthiness.

Advances in ICT provide an even stronger argument for information provenance. Groth (2004) explains that advances in computer visualisation have promoted human capabilities to recognise interesting aspects of data. How people undertake this recognition is variable and difficult to express. This results in an inability to define the discovery process and hence calls into question the integrity of the discovery. For this reason, Goth argues that an information provenance system is mandatory which tracks the human activity so that the knowledge discovery process using visualisation can be captured automatically. In a second example, Pinheiro et al. (2003) discuss the need for information provenance in the Semantic Web because more answers are derived from automatic information manipulation rather than simple information retrieval. In addition, they explain that users of information are both humans and agents and as such common sense cannot be relieved upon to judge information integrity.

Therefore, information provenance is a powerful instrument in improving information integrity. Consider this example. In the course of its enquiries a police authority collects information about an individual. This information is held within the police authority's information systems. Such information is allowed to be shared with a number of other authorised agencies across a secure network. Access is instigated by the agencies so no track is kept of where the information has been shared. Once this happens the copies of this information become legally owned by the recipient agencies. Agencies update this information for their own purposes and based upon their own intelligence. These new versions of the information are passed onto other authorised agencies. The police authority then updates the information about the individual based on new evidence. Agencies are not aware of this and continue to use their own version of the information. In this situation there exist multiple copies of the information across a complex network of agencies. Copies are not the same and there is no mechanism in place to ensure that they are the same. Clearly the integrity of the information is questionable but those receiving it are likely to be unaware of this. Decisions may be made are based on this untrustworthy information, that have detrimental effects on the individual. If the information had been accompanied by the information provenance then decision makers would be able to see how the information had changed and therefore consider how safe it was. Also, provenance would provide a method to track back to the provenance of original information held in the information systems of the police authority to check whether the original information had altered since it was first accessed.

A second example is discussed by Morozov (2009) and concerns the informational power of the internet. There are many who would argue that the internet has empowered people and allowed the free flow of information. It is a deterministic argument claiming that the introduction of the internet with its many new forms of communication, such as social networking and twitter, will increase the likelihood of democratic dialogue, freedom of speech and transparency. Morozov advocates caution. He explains that the new media not only empowers activists and human rights defenders but also nationalists, anti-democratic movements and openly extremist forces. He suggests that authoritarian governments are using new media to create propaganda via a network of "government-friendly bloggers and commentators who are paid to spin the discussion and pretend to be the voice of the people." Blog aggregators and wikis enable regimes to harvest information about emerging threats. Fake deliberative mechanisms are used to propagate unfounded confidence in growing liberalisation. This cynical information manipulation of the internet is problematic but such manipulation of the internet without provenance is both problematic and unacceptable because the questions of why, how, when, and who go unanswered.

These two examples illustrate that in the Information Society there is a moral obligation to address information integrity. Information provenance offers a normative instrument for turning this moral obligation into ethical practice.

2.4.5 Information Meltdown

We all need information all of the time. It is information that enables us to exist and flourish as humans. The converging technologies are transforming information and its access at an accelerating speed. Our traditional information bearings have been demolished which in turn threatens to send us into information meltdown. This meltdown can be avoided if we can curtail our insatiable appetite for information junk replacing it with a controlled diet of quality information in an acceptable form and delivery on demand. ICT allows us to blend forms such as prose, poetry, music and pictures and deliver them using, for example, writing and singing in either a serious or humorous fashion. We need to learn how to benefit from this varied diet of fit-for-purpose information and how to judge information is trustworthy. The Information Age offers so much but only if we master the technological keys to the informational Pandora's box.

Note: This larger section (published in 2011) includes material which also appears in 2.3 Information Provenance (published in 2005) and 2.7 Guernica Info (published in 2017). This was intentional, as it is felt that issues of information integrity and information presentation are very important and need to be revisited over time.

2.5 Trustworthy Publishing [2017]*

This section was first published as: Rogerson, S. (2016) Trustworthy Publishing. Journal of Information, Communication and Ethics in Society, Vol 14, No 1, pp. 1–3. Copyright © Emerald Publishing Limited. Reprinted by permission. DOI 10.1108/JICES-01–2016–0002

Information and communication technology has changed the nature of publishing beyond all recognition. Publishing, and academic publishing in particular, is dependent on technology—the online world has become the norm. An extensive virtual network exists in which those involved are all potentially connected. New modes and patterns of interaction and expectation have developed which challenge our social norms and moral integrity.

This network can be defined in three sections: academia, technology, and publishing. Within each section, there exists a set of relationships which must be trustworthy to be effective. Academia is founded on higher education institutions each of which comprises the institution, its academics and its students. Relationships are two-way between institution and academic, academic and student and student and institution (Rogerson, 2013). Technology deployment involves three main groups: vendors of both hardware and software; developers of both infrastructure and application; and direct and indirect recipients where direct recipients comprise clients and users and indirect recipients comprise individuals, the general public, and society as a whole (Rogerson, 2014). Once again, all possible relationships are two-way. Finally, publishing primarily comprises publisher, author, and reader with two-way relationships existing between them.

It is within this complex set of relationships that academic publishing exists. Driven by institutional strategy and priority, academics and students undertake research which is written up and submitted for publication. Technology-enabled access to data and literature informs the research and writing processes. Choice of journal, method of submission, process of review, and production of publication are similarly technology-enabled. Much of the publishing activity takes place online with many of the players never meeting face-to-face. Trusting relationships should exist throughout the publishing activity, but in the online world, this can be challenging. We are fundamentally trusting of each other—such trust is destroyed when an incident occurs that demonstrates untrustworthiness. Trusting relationships in the physical real world rely heavily on no verbal cues such as body language and tactile interaction, but in the online world, such cues rarely exist (Rogerson, 2013). Therefore, academic publishing, which is now an online activity, is exposed to the breakdown of trusting relationships.

Plagiarism, falsification, and copyright infringement are examples of causes of breakdown in trust. These issues fall within the broader field of publication ethics which is the lead topic for the first issue of Volume 14 of the *Journal of Information, Communication and Ethics in Society*. The issue begins with an invited paper, *Challenges to ethical publishing in the digital era* by Mirjam Curno. Dr Curno is a trustee and council member of the Committee on Publication Ethics (COPE). In her paper, Curno discusses three issues which illustrate vividly the challenges to the "ethical conduct of authors, reviewers and editors alike." Her call for publication ethics education of all those involved should be supported. There then follows four shorter responses to Curno's paper by well-respected academics from different fields and holding different roles.

Robert Hauptman supports Curno's position. He reminds the reader that ethical issues exist across the disciplines and as such might have different emphases and forms. His remark that data falsification might ultimately cause harm is a salutary warning. It places the act of publication in a different context. Richard Keeble's response, *Publication ethics: Stressing the positive* presents a different perspective. He suggests that "commercialisation in higher education" presents a different set of ethical challenges. He argues such challenges have their roots in the "political, social, economic and ideological fields" and as such can only be resolved in these fields. There is a sense that what is done in academia to promote ethically robust publishing needs to be backed by those in these outside fields. *The Ethics Pipeline to Academic Publishing* by Tricia Bertram Gallant supports Curno's emphasis on the role of publishing leadership. However, like Keeble, she looks beyond the publishing boundaries arguing that the urgent need is to address "systemic failure to create ethical cultures." In an ethical society, publishing ethics is likely to be accepted and flourish. Edgar Whitley provides a view from a current editor with Emerald in *Challenges to ethical publishing in the digital era: A journal editor's response to the limited mind reading skills of academic authors*. While supporting Curno's analysis, his emphasis is on unethical behaviour through ignorance. He calls for action to address this ignorance.

Whether we are authors, publishers, or readers, we have responsibilities and obligations to act in an ethically acceptable way within academic publishing. If the integrity of academic literature is to be maintained in the digital era, authors must be educated in publication ethics, publishers should actively promote ethical practice and decision makers should be mindful that strategy does not compromise ethical robustness. Only then will we have trustworthy relationships across the publishing landscape, and society will prosper.

2.6 The Data Shadow [2017]*

*This section was first published as: Rogerson, S. (2017) The Data Shadow. ACM SIGCAS Computers & Society, Vol 47, Issue 1, pp. 8–11. Copyright © the Association for Computing Machinery. Reprinted by permission. DOI 10.1145/3090222.3090225

This hypothetical story is about personal data which resides on the internet. It has its foundation in things which have happened. It raises serious questions about whether we should be more wary of, and whether there are things organisations and individuals could do to reduce the risks associated with data shadows. The paper finishes with a discussion as to how such stories might be used to educate new professionals.

2.6.1 *The Story*

The afternoon sun beat down from a cloudless sky. Her progress was momentarily recorded as her shadow loitered along behind her. It halted as she stopped to listen to the melodic birdsong. She poured cold water from her water bottle over her hair and it tickled down her neck as she tried to clear her mind.

What had gone on the night before? She could not remember—it was all an alcoholic haze. She had been woken late morning by the incessant ring tone of her mobile. It drilled into her brain like an unstoppable pneumatic drill. So many people wanting to be her friend on the social network. And all those messages from her current friends asking if it was true and if it really was her in the video. The video had gone viral—what video?—about what? She simply had no recollection at all. Her identity, once anonymous as Cassandra, had been breached—everyone knew who she really was.

Slowly she walked on. Rounding a corner, a group of teenage boys and girls looked up, pointed at her and giggled. Now she and her shadow moved on quickly trying to ignore the comments the group cried out after her. It simply would not go away—day after day after day; posted message after posted message after posted message. So much data, so many untruths accepted as fact. And then it was gone—the social network crashed. It was no longer the place to be, no longer cool, no longer where the action was. She hardly dare believe that she had her life back.

The years passed and memories of that night faded. Examination success, college graduation and a promising career beckoned. She waited patiently for her turn. Well prepared, calmly confident—this was the job she wanted—this was the job she would secure. Looking around the room the other candidates looked nervous, tense—it was hers for the taking if she kept to her game plan. Her name was called. She glided into the room, eyed the panel and smiled at them. Questions asked. Answers given assuredly. It was going well and then it happened.

Could she explain how the video of her some years earlier related and indeed supported the values she had just articulated. That video! How had it been found? She was not aware that about a year ago the founders of the social network had bought back the rights and had resurrected the "chat site with attitude" and even managed to reinstate many of the old threads, pictures and videos. They were there for anyone to access. Once again she was infamous but this time all pervasive as messages had spread across the many social networks that now existed in the online world. Self-confidence drained. The panel now seemed like both judge and jury. Dreams of a meteoric career shattered.

The afternoon sun beat down from a cloudless sky. Her progress was momentarily recorded as her shadow loitered along behind her and then it was gone forever. But not so her data shadow. It would remain permanently out there in the digital world of bits and bytes. There for all to view. There for all to judge her by.

Looking up at the sun there was a faint smile on her face and tears in her eyes. She now knew what she had to do.

2.6.2 *General Discussion POINTS*

This story raises some very important issues regarding the ethical wellbeing of personal data in the online world. For example:

- Should we be more wary of the risks associated with data shadows?
- What could organisations do to reduce the risk of inappropriate personal data existing online particularly if this existence is permanent?
- What could individuals do to safeguard the integrity of their online personal data?
- What responsibilities and obligations do computing professionals have regarding the creation and existence of data shadows?

2.6.3 *Using the Story with Undergraduates*

It has long been recognised that case studies are an effective learning tool for computer science, software engineering and information systems undergraduate students to explore the wider societal issues relating to the use of computer-based application systems. It consolidates theoretical learning through using theory to explain, act in, and solve pseudo-real-world situations. There are several ethical analysis schemes which can be used, for example, the case analysis method as defined by Bynum (2004). The tutorial is an ideal setting to undertake such explorations as it gives every student the opportunity to air their opinions in a small and safe environment. Exchange of views between students is as equally important as that between student and tutor. Experience of running many such tutorials has led to a checklist which promotes a valuable learning experience using Bynum's case analysis method.

- Brief the students to prepare the case study analysis prior to the tutorial. Students bring the completed analysis log to the tutorial as an aide memoire.
- Arrange the tutorial room so that students can all see each other.
- Move around the students encouraging participation from everyone—do not allow the more vocal student to dominate.
- Go through the following steps of the method (ibid, pp. 47–57):
  - Develop a detailed description of the case to be analysed.
  - Try to "see" the ethical issues and any "traditional" solutions that fit the case.
  - Call upon your own ethical knowledge and skills.
    - Take advantage of one or more systematic analysis techniques such as "ethical-theory analysis" or "professional standards analysis."
  - Draw relevant ethical conclusions about the case.
  - Draw relevant lessons about the future.
- Encourage the students to put forward their own ideas rather than the tutor putting forward his/her own views.
- Act as the arbitrator and guide.

This process can be completed in a one-hour tutorial with a maximum of 16 students. There are many case stories which can be used in this learning schema. Some may be fictitious but rooted in the possible, as is the previous story. Two further examples written by the author are *GARAGE auction website* (Rogerson, 2016a) and *Nancy's ethical dilemma* (Rogerson, 2016b). Some may be real stories of ethically charged events. The Volkswagen emissions scandal and Tesla Autopilot software are examples recent examples (Rogerson, 2017).

2.6.4 And Finally . . .

Society at large needs competent, empathetic and altruistic professionals to deliver societally-acceptable fit-for-purpose systems. The education of new professionals is paramount. Teaching technology in isolation is unacceptable and dangerous.

> The opportunity to participate in an active rather than passive manner leads to an experiential journey of maturity from tutor-led activities to student-led activities. Through this process, ICT professionals of the future are more likely to gain the necessary skills and knowledge to act in a socially responsible manner not on the basis of instinct and anecdote but on rigour and justification.

(Rogerson, 2015)

2.7 Guernica Info [2017]*

This section was first published as: Rogerson, S. (2017) Guernica Info. In: Rogerson, S. Ensalada Mixta: Observations and Inspirations, pp. 41–45, ISBN 978-1-9998826-0-0. Copyright © Simon Rogerson.

Guernica is a compelling story which holds many messages and lessons for every following generation. I have used the story to illustrate how information can be conveyed to people about catastrophic events which have the potential to affect everyone's life.

2.7.1 A Brief Historical Account

On 26 April 1937 the Basque town of Guernica suffered the first carpet bombing of a civilian population in times of conflict. It was carried out by the German air force's Condor Legion and the Italian Aviazione Legionaria in support of the Spanish nationalist government. The attack on the town took place without warning or obvious provocation.

It was market day and the town's population of 7,000 had swollen to 10,000. At 4.30 in the afternoon the first of many waves of raids occurred. The town was bombed and strafed for three hours. The number of dead and wounded civilians was horrendous. The town was left in ruins.

Many fictional and non-fictional accounts have been written about the Spanish Civil War and the atrocities which occurred. The most moving I have found to be: *Guernica* by Dave Boling; *Guerra* by Jason Webster; *Nada* by Carmen Laforet and translated by Edith Grossman; and *Soldiers of Salamis* by Javier Cercas and translated by Anne McLean.

2.7.2 Telling the Guernica Story

One of the most stunning forms of information is Picasso's Guernica which he painted in 1937 because he was so outraged by the atrocity. The painting is in black and white without colour and measures 3.5m by 8m. Using IT, it is possible to project this image and discuss the events of Guernica against a backdrop of the sound of bombing. There have been many critiques of Picasso's painting. Often lengthy and detailed they each interpret the painting searching out the intrinsic messages and meanings. Displaying these descriptions provides a sense of gravitas through their sheer length before finishing with a one-sentence summary, "A composition so compelling challenges our most basic notions of war as heroic, unmasking it as a brutal act of self-destruction."

Another way to tell the story is through picture and song. A series of black and white photographs are projected whilst playing *Market Day in Guernica* sung by Katie Melua and written by Mike Batt. The pictures are sequenced for maximum impact as the lines of the song unfold; . . . My little ones no longer play In Guernica on market day . . . I watched as he was blown away In Guernica on market day. . . . This IT-enabled portrayal always provokes emotional reaction.

Information currency and value are vividly illustrated by Picasso's painting. A tapestry copy hangs on the wall of the Headquarters of the United Nations in New York, at the entrance to the Security Council room as a reminder of the horrors of war. On 5 February 2003, Colin Powell held a press conference at the UN to justify the war on Iraq. It was claimed that the televisions news crews had asked for a blue cloth to be placed over the tapestry because it interfered with the image of the speaker. It was later discovered that it was the Bush Administration which had pressurised UN officials to cover the anti-war tapestry. Visual impact during presentations is used to emphasise this political wrongdoing by projecting first *Guernica cover up [2003]* by the artist Neil Jenkins. This is a blue tapestry copy of the painting to represent the cover up. A small part of the tapestry is then enlarged to reveal the images in the tapestry have been embroidered using the words of Colin Powell's speech. This never fails to surprise the audience. In 2016 Syria's civil war escalated with the death of thousands of civilians. The carpet bombing of Aleppo has been likened to Guernica. Once more the messages of Guernica are current and apposite.

A new way of exploring Guernica is through using augmented reality. Augmented reality supplements the view of the physical real world with sound, video, graphics or GPS data. It is claimed that IT thus enhances one's current perception of reality. A software App for mobile phones, iPads and tablets called Augmented Guernica provides a 3D interpretation of the painting. Pointing the camera at a picture of the painting in a book or even the painting itself turns the painting into a three-dimensional image which can be explored in many different ways.

2.7.3 The Story's Impact

The story, sights, and sounds relating to Guernica are new and shocking for many in every audience where this has been presented. Reactions vary to the different forms of presentation, but the least impact is always the textual form. The use of IT to create, communicate and consume information can add and subtract value from its original form but such change is unlikely to be uniform across those involved. This is simply because people's perceptions are different.

References

Borgmann, A. (1999). *Holding on to reality: The nature of information at the turn of the millennium*. University of Chicago Press.

Brey, P. (2006). Evaluating the social and cultural implications of the internet. *Computers and Society, 36*(3), 41–48.

Bynum, T. W. (2004). Ethical decision-making and case analysis in computer ethics. In T. W. Bynum & S. Rogerson (Eds.), *Computer ethics and professional responsibility* (Chapter 3, pp. 60–86). Blackwell Publishing.

Eaton, J. J., & Bawden, D. (1991). What kind of resource is information? *International Journal of Information Management, 11*(2), 156–165.

Evans, C. (1979). *The mighty micro: The impact of the computer revolution*. Victor Gollancz Ltd.

Fairweather, N. B., & Rogerson, S. (2003). The problems of global cultural homogenisation in a technologically dependant world. *Information, Communication & Ethics in Society, 1*(1), 7–12.

Fox, M. S., & Huang, J. (2003). Knowledge provenance: An approach to modeling and maintaining the evolution and validity of knowledge. In *EIL Tech. Rep.* Enterprise Integr. Lab., University of Toronto, Toronto, ON, Canada.

Groth, D. P. (2004). Information provenance and the knowledge rediscovery problem. In *Proceedings of the eighth international conference on information visualization* (pp. 345–351). IEEE Computer Society.

Hongladarom, S., & Ess, C. (Eds.). (2007). *Information technology ethics: Cultural perspectives*. Idea Group Inc.

Huang, J., & Fox, M. S. (2004). Uncertainty in knowledge provenance. In *Proceedings of the 1st European semantic web symposium*. Heraklion, Greece.

Macgregor, G. (2005). The nature of information in the twenty-first century: Conundrums for the informatics community? *Library Review, 54*(1), 10–23.

Mandke, V. V., & Nayar, M. K. (2004). Beyond quality: The information integrity imperative. *Total Quality Management and Business Excellence, 15*(5–6), 645–654.

McRobb, S., & Rogerson, S. (2004). Are they really listening? An investigation into published online privacy policies. *Information Technology and People, 17*(4), 442–461.

Meyer, H. (2005). The nature of information and its effective use in rural development. *Information Research, 10*(2). Retrieved October 20, 2006, from http://informationr.net/ir/10-2/paper214.html.

Moreau, L., Groth, P., Miles, S., Vazquez-Salceda, J., Ibbotson, J., Jiang, S., Munroe, S., Rana, O., Schreiber. A., Tan, V., & Varga, L. (2008). The provenance of electronic data. *Communications of the ACM, 51*(4), 52–58.

Morozov, E. (2009). Censoring cyberspace. *RSA Journal, 155*(5539), 20–23.

Nance, K. L., & Strohmaier, M. (1994). Ethical accountability in the cyberspace. In *ECA94: Proceedings of the conference on ethics in the computer age* (pp. 115–118). Association for Computing Machinery.

Nielsen J. (2003, October 13). Web guru fights info pollution. *BBC News Online*. Retrieved October 24, 2006, from http://news.bbc.co.uk/2/hi/technology/3171376.stm.

Pinheiro da Silva P., Deborah, S., McGuinness D. L., & McCool, R. (2003). Knowledge Provenance Infrastructure. *IEEE Data Engineering Bulletin, 26*(4), 26–32.

Rogerson, S. (2013). The integrity of creating, communicating and consuming information online in the context of higher education institutions. In L. Engwall & P. Scott (Eds.), *Trust in universities, 86* (Chapter 10, pp. 125–136). Retrieved January 8, 2016, from www.portlandpress.com/pp/books/online/wg86/086/0125/0860125.pdf.

Rogerson, S. (2014, April). Preparing IT Professionals of the Future! *Mondo Digitale, 50*(2), AICA—Associazione Italiana per l'Informatica ed il Calcolo Automatico.

Rogerson, S. (2015). Future vision. *Special Issue—20 years of ETHICOMP, Journal of Information, Communication and Ethics in Society, 13*(3/4), 346–360.

Rogerson, S. (2016a). *GARAGE auction website*. Retrieved February 11, 2017, from www.linkedin.com/pulse/garage-auction-website-simon-rogerson?trk=mp-author-card

Rogerson, S, (2016b). *Nancy's ethical dilemma*. Retrieved February 11, 2017, from www.linkedin.com/pulse/nancys-ethical-dilemma-simon-rogerson?trk=mp-author-card.

Rogerson, S. (2017). Is professional practice at risk following the Volkswagen and TESLA motors revelations? *The ORBIT Journal, 1*(1), 1–15.

Rogerson, S., & Bynum, T. W. (1995). Cyberspace: The ethical frontier. *The Times Higher Education Supplement, 1179*(9), 4.

Statista (2021, June 7). *Volume of data/information created, captured, copied, and consumed worldwide from 2010 to 2025*. Retrieved October 23, 2021, from www.statista.com/statistics/871513/worldwide-data-created/.

Toffler, A. (1970). *Future shock*. New York: Bantam Books.

Chapter 3

Systems and Applications

The length of this chapter demonstrates the scope of digital technology application. Many of the major application areas since 1995 are covered in this chapter. The broader issues are teased out alerting the reader to the need for ethical caution when creating and implementing application systems. In the following table, the 31 sections have been grouped together according to their primary focus and in chronological order within each grouping.

Primary focus	Section	Title	Original publication date
Access to services	3.4	Smart Cards	1998
Access to services	3.12	At Cross Purposes	2002
Access to services	3.15	Playing Your Cards Right	2003
Automation	3.28	Ethical Robots	2011
Healthcare	3.8	Keeping Records Safe	2000
Healthcare	3.9	Health Warning	2001
Healthcare	3.27	The Way to Healthy ICT?	2010
Healthcare	3.31	Hospital Safety	2021
Manufacturing	3.2	Making BPR a Success	1996
Online social media	3.1	Information Superhighway	1995
Online social media	3.3	People Issues	1998
Online social media	3.17	What is Wrong with Mobile Phones?	2003
Online social media	3.21	Let the Games Begin	2006

(Continued)

DOI: 10.1201/9781003309079-3

(Continued)

Primary focus	Section	Title	Original publication date
Online social media	3.22	Ethics of the Blogosphere	2006
Online social media	3.23	Voicing Concern	2008
Online social media	3.26	The Social Side of Wireless	2009
Online social media	3.30	People are Data in the Connected World	2018
Privacy	3.5	Surveillance	1998
Privacy	3.7	Identity Crisis Online	2000
Privacy	3.10	Facing Up to the Issues	2002
Privacy	3.13	Matching Cases	2002
Privacy	3.19	Tag Ethics	2004
Privacy	3.24	Ethics of the Street	2009
Privacy	3.25	Ensuring Ethical Insurance	2009
Retail and trading	3.6	Electronic Commerce	1998
Retail and trading	3.14	Going, Going . . . Gone!	2003
Retail and trading	3.18	Market~ing Forces	2004
Safety and security	3.16	Trusting the System	2003
Safety and security	3.20	A Lesson from History	2006
Software development	3.11	Clear and Present Danger	2002
Software development	3.29	Ethics Omission Increases Gases Emission	2018

As digital technology becomes more pervasive, so the range of applications becomes wider. In part this is driven by online social media which has expanded development contribution, user access and global coverage. It is therefore unsurprising that there are eight sections which have online social media as the primary focus; the first being published in 1998 and the last in 2018. It also follows that privacy of individuals was and remains a central ethical issue for digital technology; while seven sections have this as the primary focus, many others also include privacy as a secondary issue.

3.1 Information Superhighway [1995]*

This section was first published as: Rogerson, S. (1995) ETHIcol—information superhighway. IDPM Journal, Vol 5 No 4, p. 23. Copyright © Simon Rogerson.

Even the most casual observer of social trends and workplace trends will have heard of the amazing information revolution that is likely to affect, directly or indirectly, every citizen in the world. The revolution is referred to as the information superhighway, or the internet, or the world-wide web, or cyberspace. This is a global network of seemingly boundless information presented as text, pictures, graphical images, sound and video that can be accessed easily and cheaply by any individual. It is a network of millions of connections that is expanding at 20% per month! Virtually every day, the media carries news about the superhighway. It may be concerning a technological advancement but more likely it is a story about the superhighway's effect on the life of an ordinary individual. Those who are not conversant with the superhighway will become more and more disadvantaged. Those who believe it is a passing fad are sadly out of touch. Virtually all large organisations now have a presence in cyberspace. This is a revolution probably more significant than the printing press and the internal combustion engine. Why?—because of its meteoric pace, its breadth of impact and its removal of temporal and geographical barriers.

Without doubt, the ability to disseminate information quickly and effectively to many people and the ability to work with appropriate individuals regardless of their physical location has many benefits. For example, a leading surgeon performing life-saving operations on patients thousands of miles away is now reality. There are, however, problematical issues that everybody, and particularly computer professionals need to be aware of.

E-mail is an established form of communication that has delivered real benefits particularly when linking to external mailboxes via the internet. Subscribing to specialist interest groups can provide a rich seam of information and idea generation but there are pitfalls. There are increasing incidents of antisocial behaviour called spamming where an unsolicited message, often detailing a product or service, is sent to several thousands of lists and thus many thousands of individuals. Such activity clogs up communication resources to the point where legitimate messages may fail to be delivered. A recent example involved an offer of a one-year free subscription to a magazine with the hidden agenda of enticement to purchase other goods and the purloining of personal data. There was an outrage and a hunt was instigated to find the sender. At the time of writing, the culprit is close to being caught. The progress of the hunt invoked as much mail as the offending message, a knock-on effect that seemed to go unnoticed. Redress to the law in these cases is very difficult due to the global nature of the problem. In this lawless society individuals seem to be driven to applauding the cyberspace bounty hunters whose motives seem to be an unholy alliance between a craving for notoriety and a desire to do good.

The Archbishop of York recently expressed concern over the effects of becoming obsessed with violent and obscene material available on the internet. Indeed, this is a fast-growing area of activity with pornographic material being easily accessible to anyone, including the young, who has the ability to log onto the internet. A particularly worrying trend is the exploitation of vulnerable groups. For example, an e-mail list for disabled students was recently subjected to offers of erotic material for individuals and partners. Violent groups, such as neo-Nazis, have used the internet to promote their particular brand of rhetoric and violence.

Turning to the issue of social groupings, the superhighways are having some potentially damaging effects. Teleworking is reliant on the superhighways. Whilst teleworking can be very effective for organisations, it can isolate individuals. A recent television documentary depicted a marketing executive of an international firm as quite a pathetic figure, teleworking across the globe but looking forward to the social interaction with the postman to break the monotony and chill of cyberspace. Commercial opportunists have moved cyberspace into the high street with cyber pubs and cafes where the fixed menu includes food, drink, and net surfing. Just as video arcade games can become obsessional so can surfing the net. Conversation is replaced with the tapping of keyboards and the glazed stare of the surfing addict.

It might appear to be an overstated bleak picture but these problems are occurring and are increasing. It is the responsibility of every professional to help ensure that this powerful enabling technology does not turn into an enslaving one.

3.2 Making BPR a Success [1996]*

This section was first published as: Rogerson, S. (1996) ETHIcol—making BPR a success. IDPM Journal, Vol 6 No 2, p. 23. Copyright © Simon Rogerson.

Much has been written about the success and failure of Business Process Re-engineering (BPR). Leadership, organisational, cultural, and people issues have been identified as the major obstacles in achieving BPR success. Without doubt, there will be winners and losers in this activity. There is evidence to suggest that BPR errs on being socially insensitive and perhaps this is why most re-engineering efforts have had little measurable impact on the overall business. Consequently, it is important that organisations about to embark on such programmes are fully aware of the potential impact on individuals, groups, and society as a whole.

Given the major impact that BPR has on an organisation and that this impact involves the generation, dissemination and use of information to sustain the redesigned processes it is inevitable that IT has a central role in this activity. For example, telecommunications often figures in reducing co-ordination costs or increasing the scope of co-ordination, and shared databases are commonplace in the provision of information across and during processes.

Within BPR there are numerous activities and decisions to be made and most of these will have an ethical dimension. There are issues relating to both the process and the outcome of BPR. In his recent book, *The Re-engineering Revolution—The Handbook*, Michael Hammer sets out ten guidelines based on potential reasons for failure. The two most ethically significant are:

- Make sure that you know what re-engineering really is before you attempt to do it and then do it, not something else.
- Any successful re-engineering effort must take into account the personal needs of the individuals it will affect. The new process must offer some benefit to the people who are, after all, being asked to embrace enormous change, and the transition from the old process to the new one must be made with great sensitivity to their feelings.

Opinion recently canvassed from BPR practitioners and employees affected by BPR initiatives located in the UK, US, the Netherlands and Sweden revealed a number of ethical issues which included:

- the damaging of trust between employer and employees through insensitive BPR
- the existence of hidden agendas such as downsizing
- censorship on full disclosure
- conflicts between what the BPR client wants or expects compared with what is considered to be correct action

The following case study demonstrates what can go wrong. A large service-based organisation in the UK realised its ongoing programme to expand through opening regional offices in light of increased demand was not financially sustainable and indeed would put the company in a perilous state. It decided to re-engineer and hired an independent consultant to co-ordinate this activity.

The CEO wanted to reduce the workforce and to move to a home-based operation where those directly involved with clients worked from home rather than from regional offices as these were costly to maintain. Communication about the re-engineering project tended to be one-way pronouncements from the top. Reassurances were given about job security at an early stage. Processes were redesigned but not piloted. IT was specifically excluded from the BPR exercise, a decision being made to consider this after the re-engineering dust had settled. The result of the BPR exercise was a significant reduction of the company's middle management, the closure of some regional offices with associated job losses, the move to more home-based working, and the majority of staff had to reapply for their jobs most of which had been modified. These major changes, which were very upsetting for many people, where announced without any prior warning. There then followed several months between the announcement and implementation, a time when all staff were expected to work towards establishing the new style of organisation. The effect of this BPR exercise was a loss of loyalty and motivation in the workforce, a high turnover of staff who were identified as being key in the new structure and 18 months later the organisation had to embark on drastic remedial action.

But it does not have to be like this. Corning is the biggest company in the smallest town in the US, with half the town's 12,000 inhabitants working for the company. In this environment, the company has a long track record of social responsibility. The company decided to re-engineer after its market value plunged by 25% and significant losses were announced. It made it clear from the onset that it would not re-engineer the company at the detriment of the town. Corning got the workers to do their own re-engineering, releasing key employees from their normal jobs for several months in order to undertake the BPR exercise. Easy access to senior management was established and there was a lot of effort put into informing everyone of progress and enabling frank discussion to occur. The outcome was focused on reduction of complexity in the corporate structure, the spreading of best practice, evolutionary change rather than revolutionary change, and even-handedness regarding job losses with an emphasis on early retirement rather than redundancy. The following year, Corning returned to profit and sales increased by 19%. It will take several years before BPR has its full effect but it appears from this evidence that Corning has succeeded in turning the company round. Social responsibility is very visible in the approach that Corning adopted.

Organisations considering undertaking BPR as well as those in the process of a BPR exercise must temper their actions to ensure that BPR is ethically sensitive thus increasing the probability of achieving genuine success for the greatest number of people. To encourage this there should be:

- explicit consideration of ethical and social responsibility in the BPR process
- training of BPR exponents in ethical practice to improve implementation success
- adherence to ethical principles by BPR exponents

3.3 People Issues [1998]*

*This section was first published as: Rogerson, S. (1998) ETHIcol—People Issues. IMIS Journal, Vol 8 No 4, p. 9. Copyright © Simon Rogerson.

In the Information Society we must not forget that people matter. How people present themselves, react to the new stimuli of cyberspace and converse with those from other cultures will have a dramatic effect. The issues of identity, culture and conversation illustrate this point.

3.3.1 Identity

There has been much speculation that people can play with their identities on the internet and present different electronic persona. This might be possible in the short-term but such mimicry is likely to be spotted in the long term as identity is more than simply learning and applying rules, it is about learning within a context and reacting intuitively to different situations as they arise. In the Information Society, the impact of physical characteristics of those communicating has been minimised. This is potentially beneficial as removing the visual cues about gender, age, ethnicity and social status allow different lines of communication to open up that might have been avoided in the physical world.

3.3.2 Culture

The Information Society crosses traditional boundaries and as such comprises individuals from many different cultures. This cultural variability means that the expectations of individual cyber-citizens can differ considerably. There are two important dimensions to consider regarding this variability (see Section 2.4). The first dimension is the continuum from individualism to collectivism. Individualism emphasises self-interest and promotes the self-realisation of talent and potential. Its demands are universal. Collectivism emphasises pursuit of common interests and belonging to a set of hierarchical groups where, for example, the family group might be placed above the job group. The demands on group members are different to those on non-group members. The second dimension concerns cultural differences in communication. In low context communication the majority of information resides in the message itself whilst in high context communication is implicit. It has been suggested that the US utilises low context communication whilst Japan uses high context. Given such cultural variability it is clear that there are great difficulties in achieving an internet that is acceptable to all.

3.3.3 Conversation

There is concern by psychologists that computers are having a detrimental effect on society in that the "social glue" of casual conversation is being eroded. The increasing use of email at work, the elimination of bank tellers and shop assistants, and the use of telephones and laptops to telecommute from the home illustrate how the opportunity for small talk is decreasing. This phenomenon is thought to be one reason why shyness is increasing in the population. There is of course a counter argument in that by utilising computer based communication tools such as the internet those who are naturally shy become more outgoing since the psychological pressure of face to face contact is removed.

3.3.4 One Planet, One Net—Principles for the Internet Era

In the US, Computer Professionals for Social Responsibility (CPSR) has been addressing how the internet might be administered to ensure that it evolves for the benefit of all humanity. CPSR recognises that the internet comprises the collective knowledge and experience of countless communities, each with its own modes of interaction, languages of discourse, and forms of cultural expression. This has led to a proposal for the adoption of seven principles that must be understood and respected if the internet is to evolve in an acceptable manner. The principles are as follows (extracts from the original CPSR document):

1. **The Net links us all together.** The nature of people and their use of networking technology provides a strong natural drive towards universal interconnection.
2. **The Net must be open and available to all.** The Net should be available to all who wish to use it, regardless of economic, social, political, linguistic, or cultural differences or abilities.
3. **Net users have the right to communicate.** Every use of the Net is inherently an exercise of freedom of speech, to be restricted only at great peril to human liberty. The right to communicate includes the right to participate in communication through interacting, organising, petitioning, mobilising, assembling, collaborating, buying and selling, sharing, and publishing.
4. **Net users have the right to privacy.** Without assurances of appropriate privacy, users of the Net will not communicate and participate in a meaningful manner.
5. **People are the Net's stewards, not its owners.** Those who want to reap the benefits of the shared global Net are obliged to respect the rights of others who may wish to use the Net in different ways. We must work to preserve the free and open nature of the current internet as a fragile resource that must be enriched and passed on to our children. Just as the ecosystem in which we live cannot be owned, the Net itself is not owned by anyone.
6. **Administration of the Net should be open and inclusive.** The Net should be administered in an open, inclusive, and democratic manner for the betterment of humanity. The needs of all who are affected by the internet—including current users, future users, and those who are unable to or choose not to be users—must be considered when making technical, social, political, and economic decisions regarding the operations of the internet.
7. **The Net should reflect human diversity, not homogenise it.** The Net has the potential to be as varied and multi-cultural as life itself. It can facilitate dialogue between communities and individuals that might previously not have encountered each. In order to preserve the vitality that comes with a diversity of viewpoints, we should work toward helping the whole world participate as equals.

Clearly there will be different suggestions as to how the internet should evolve. An Information Society that empowers the disabled and less fortunate members of society and sustains equality of opportunity regardless of race, colour, or creed is achievable. The policy makers, developers and service providers of the Information Society have the wherewithal to create to make this happen but they must have the commitment as well else disaster beckons.

3.4 Smart Cards [1998]*

This section was first published as: Rogerson, S. (1998) ETHIcol—smart cards. IMIS Journal, Vol 8 No 1, p. 33. Copyright © Simon Rogerson.

The smart card is one of the digital icons of the Information Age. Smart card technology is being applied in various ways to facilitate trade, gain access to services and products, verify identity, and establish and influence relationships. In the UK there have been many applications, for example, the electronic purse—Mondex, the Shell loyalty card and the Social Security Benefits Card. Similar examples can be found in different parts of the world. In Spain a smart card has been introduced for benefit payments and access to government databases. A smart patient data card is being tested in a region of the Czech Republic to replace the paper-based system that had limited capacity, was inaccurate, labour intensive to maintain and open to widespread abuse. Two million smart cards have been issued to the poor in Mexico for distributing food and cash benefits.

A recent study found that 27% of smart card applications were within banking, 18% within health and welfare and 15% within transport. Other applications included; telecommunications, identification, phonecards, retail loyalty schemes, metering, radio security, physical access, and gambling. The use of multifunctional smart cards was commonplace.

Smart cards have three broad functions; authentication, storing value and storing personalised information. Authentication is concerned with ensuring only authorised individuals gain access to systems and buildings. A smart card can be used as an electronic purse to store units of value in different currency denominations as well as credit and other units of value such as bonus points or air miles. Values can be replenished on a smart card. The smart card can also be used as a portable storage device independent of some fixed location and with the capability of holding a large amount of data of different forms and for different purposes but usually of a personal nature.

Clearly there are beneficial outcomes from the application of smart cards. Realising these benefits both for individuals and organisations may well profoundly change the relationship between clients or consumers and suppliers or government bodies. A smart card that is your passport, driving licence, credit and debit card, access to your place of work and your car ignition key will undoubtedly alter relationships due to potential uneasiness about what data is held, accessed and modified. Such cards are already being piloted. For example, in South Korea a national citizen card is being introduced which is used as a driving licence, identity card, pension card, and medical insurance card.

Some of the potential benefits of smart cards are:

■ using smart cards is safer than carrying cash for an individual
■ smart cards can improve access to services for the disabled and elderly
■ it is a secure means of authenticating one's own identity
■ it is a secure means of authenticating the identity of reader device
■ it is a portable and secure store of information available to all
■ access can be made available in geographical locations where on-line communication is not possible
■ the opportunity of fraud is reduced using smart cards
■ social disadvantaged groups can gain access to facilities and resources without feeling stigmatised
■ objective selection criteria can be upheld and the risk of bias or favouritism reduced

Consider just one example. Smart card technology has the capability of addressing access, independence and equality of opportunity issues for the disabled through facilitating adaptive interfaces. Individual requirements could be stored on the smart card so that the interface at the point of use would automatically adapt to the preferred customer verification method (for example hand geometry), input (for example voice activation and speech recognition), operation (for example reduced functionality), and output (for example large colour specific characters). Contactless smart cards could be used to remove the necessity of card insertion into readers, to unlock and open doors, to activate location signals, to increase road crossing times and to adjust access heights of facilities.

There are potential pitfalls for individuals and society in general regarding smart card applications and these include:

■ smart cards lead to a loss of anonymity
■ pseudonymity can be mistaken for anonymity as card schemes indirectly hold cardholder identity

- smart card schemes could lead to a reduction in the provision of non-smart card facilities and so affect freedom of choice
- smart cards can reduce access to services and resources for the technology illiterate or technology wary
- there are difficulties in viewing personal data by card holders
- smart cards can result in significant invasions of privacy
- profiling and tracking of individuals can occur
- increases in smart card use could lead to a de facto national and subsequently global identity card that has not been subjected to citizen consultation
- smart card functionality can be increased without proper consideration of the overall impact

It has been suggested that a number of principles should be adhered to when considering if and how a smart card scheme should be implemented. Of these the key principles are:

- smart cards must properly respect the legal and ethical rules pertaining to the rights of the card holder
- individuals should have the right to refuse a cards
- the card holder's prior consent is required for all uses of the card and disclosure of information it contains
- cards should not be used as tools for overt or covert surveillance

Having decided to implement a smart card scheme certain design features seem appropriate and are summarised as:

- Identified transaction trails should only be used where no acceptable alternative exists.
- Identity should be safeguarded using pseudonymity.
- Ensure integrity across applications on multi-purpose cards.
- The design of smart card schemes must be transparent to the individual.
- Biometric and encryption key data should be held on the card.
- Two-way device authentication must be used.

Smart cards offer great potential benefits to society. Given its pervasive nature careful policy, design and implementation strategies must be in place. With these, one can envisage a time when the lack of ownership of a multi-functional smart card will result in a dramatic loss of opportunity and of help in times of need for the "non-citizen." The aim must be to achieve sensitive usage and ensure ordinary people are involved in the technological decision-making process which precedes application of smart card technology.

3.5 Surveillance [1998]*

This section was first published as: Rogerson, S. & Fairweather, N.B. (1998) ETHIcol—surveillance. IMIS Journal, Vol 8 No 3, pp. 15–16. Copyright © Simon Rogerson & N Ben Fairweather.

In this edition we consider some of the issues associated with electronic surveillance in the workplace. While there is no doubt that employers have a legitimate interest in a certain amount of monitoring of their employees, to ascertain employee efficiency and effectiveness, there is also no doubt that employees are not slaves and as such should not be required to reveal their whole selves to their employer. To allow intimate information to remain private, both traditional workers and teleworkers should not normally have personal communications under surveillance by their employer, and the employer should not routinely monitor the length of time employees spends away from their work during the working period.

Over the last few years, software packages have made electronic performance monitoring possible. This significantly expands the scope of management to collect and process data on their workers. Monitoring will be more intense when workers use, for example, e-mail, video conferencing and dedicated on-line systems to carry out their work.

The practice of monitoring workers' communications is widespread. Richard Spinello suggests that it is carried out in a covert fashion by a growing number of firms such as Pillsbury, UPS, and Intel Corporation. According to The New York Times others, for example, Kmart Corporation have adopted a policy that allows the company to review all e-mail messages, and every employee is informed of this policy at orientation meetings.

Fear that communication is being monitored may make employees less likely to use company resources for personal business. However, in a society where many employees work full-time and many live alone there may be no choice but to conduct some personal business from the workplace. This need is especially strong for those working conventional hours as these are likely to be the only opening hours of many of the offices that they need to deal with in their private lives. In such circumstances, it is unreasonable for an employer to expect that no use at all is made of company time and resources for private business that remains private. However, when use for private business is significantly detrimental to the performance of the job that the employer has paid for, or uses a valuable quantity of resources, it is reasonable for an employer to take action, which might include monitoring.

Whilst employers might argue that electronic monitoring acts as an effective deterrent against fraud, industrial espionage and other illegal activities this does not give them a universal right to monitor their employees. Most of us would not consider it acceptable if all of our post were read, which is why there are laws against such interception of post. Just because there is the possibility that someone might abuse e-mail does not mean that everyone's e-mail should be monitored. The civil liberties of innocent people should not be suppressed because a few rogue employees might abuse those liberties. If a company has legitimate suspicions that someone is using its e-mail system for illegal reasons, it should investigate and take any necessary action. A breach of privacy would surely be justified under such circumstances as long as there are appropriate procedural safeguards. When there are no such suspicions, the possibility of such abuse of the systems should not outweigh the reasonable expectation of employees to be trusted by their employer.

On 25 June 1997, the European Court of Human Rights ruled that workers have a "reasonable expectation" of privacy in making and receiving telephone calls at work. It is reasonable to assume that this applies equally to e-mails and other communications at work. Some organisations contend that because they own the computer resources on which e-mail messages are transmitted, they should have an unconditional right to control and monitor the contents of those messages. The European Court of Human Rights was quite clear that ownership does not permit surveillance, in rejecting the UK Government's case that taps of telephones were not in breach of human rights where telephones were government property.

Monitoring is particularly likely with e-mail because unlike other forms of communication e-mail provides an easy means of saving and inspecting messages for many years. With other forms of communication, the setting up of recording and monitoring may be relatively complex and time consuming. With e-mail messages already in ASCII code, using a computer to search large numbers of e-mails for mentions of particular words is as easy as searching in a document being word-processed.

One worry with all types of surveillance and monitoring is that they give power over those monitored to those who carry out the monitoring. This may be greater than the power it is appropriate for them to have, especially when they come to know intimate information about those monitored. This can cause a variety of problems, including breaches of internal security and giving knowledge that amounts to organisational power to those who do not deserve it.

The fear that intimate or organisational information may fall into the wrong hands may lead those who are subject to surveillance or monitoring to be constantly apprehensive and inhibited in actions and communication that may be monitored. In such circumstances employees may feel they are under surveillance of a kind they resent more than traditional workplace monitoring. Where technological methods have been introduced for such detailed monitoring or surveillance, they have in some cases been associated with very low employee morale.

Peer-to-peer communication always exists within any work situation. Business and personal matters become interspersed as workers inadvertently discuss aspects of their personal lives as they conduct business. Such interactions have the potential to make even the most boring jobs bearable. Furthermore, such informal communication can provide the breeding ground for ideas about how work processes could be improved and waste reduced.

If such peer-to-peer communication is facilitated through video conferencing, e-mail or telephone calls at the employer's expense, there may be a strong potential for surveillance of such peer-to-peer communication. Managers may have a fear of such communication, as it enables workers to share stories detrimental to the image of the manager, or co-ordinate complaints. Equally, such time spent not at the primary work task could be seen by managers as wasted time. Such concerns may lead managers to pay particular attention to peer-to-peer communications. Such action might stifle team building and valuable information exchange, and ultimately might lead to a reticent workforce.

Most would agree that some workplace monitoring is a legitimate and essential management activity. However, there is a suspicion that close technological monitoring takes place more because it is technologically easy rather than because there is any great need for it. It is then that monitoring crosses the line into unethical business practice.

3.6 Electronic Commerce [1998]*

This section was first published as: Rogerson, S. (1998) ETHIcol—electronic commerce. IMIS Journal, Vol 8 No 5, p. 13. Copyright © Simon Rogerson.

Electronic commerce in many ways is the ultimate symbol of the Information Age. Trading is a fundamental activity of even the most primitive of societies and so it was inevitable that as more human activities moved into the virtual world so the ability to trade virtually became essential. Electronic commerce is about delivering information, products, services, and payments via electronic media where the supplier and customer are physically remote.

Electronic commerce may become the dominant form of trading in the next millennium. In a survey undertaken for IBM, 71% of IT directors reported that electronic commerce was a key element of their organisation's business strategy and 82% believed that failure to embrace electronic commerce within two years will put an organisation at risk. The Yankee Group research organisation predicts that internet-based retail sales will grow from $9.7bn in 1997 to $96bn in 1999, while others predict electronic commerce will be a $1.5 trillion industry by 2000. The top 100 UK companies believe that 20% of their revenue will come from electronic commerce by 2000 according to the KPMG Electronic Commerce Research Report in the UK in 1997.

The claim of governments is that electronic commerce, whether business to business or business to consumer, is not just a new medium for selling but a new market that will contribute to competitiveness, growth and jobs. Whilst this might be true and that these facilities certainly offer benefits to, for example, the housebound or those who work on shifts, it is far from clear whether the implications of electronic commerce are understood. It may be that certain aspects are potentially detrimental and should be modified. Generally electronic commerce raises a number of issues regarding, for example, privacy, security and freedom of speech.

Overall electronic commerce does not appear fundamentally different to traditional trading. Where differences do occur is in the manner in which activities are undertaken. This appears to present new ethical challenges. Many activities in traditional shopping can occur without consumer identity being divulged. This is not so with electronic commerce as increasingly detailed identification is required as the trade progresses. The manner in which identification takes place is important as there is potential for traders to be unacceptably intrusive when dealing with consumers. There is no choice in whether identity is to be withheld or divulged in the electronic trade which certainly impinges on consumer autonomy.

In purchasing a physical product electronically, a lot of information about the consumer is collected by the supplier including choice of product, delivery address, number, and times of electronic shopping visits, amount spent, and payment information. Clearly it is possible and inevitable that the supplier will save this information and build up customer profiles for future use. It is also possible that this information is passed to third parties such as credit agencies, tax agencies and marketing companies. The cookie is an example of an information collecting agent used in electronic commerce by host websites to record information about their visitors. This agent is highly intrusive and mostly used surreptitiously. To address consumer concerns, it has been suggested that web pages should contain warnings that cookies are in use, that consumers should be able to suppress cookies and that consumer consent, whether implicit or explicit, should be obtained prior to cookies being invoked.

Traditional trading relies heavily upon nonverbal communication. In electronic commerce the impact of physical characteristics of those communicating has been minimised in all activities. This is potentially beneficial as removing the visual cues about gender, age, ethnicity and social status allows different lines of communication to open up that might have been avoided in the physical world and this might lead to fairer and more open trading. However, the removal of nonverbal communication means that trading is different. Gone are the intuitive signals gleaned from body language, gone is the physical token passing and gone is the handshake to close a deal. Trading is now solely dependent on the electronic word which is susceptible to distortion and misunderstanding particularly if trader and consumer are from different countries or cultures. This may well increase the risk of more vulnerable members of society being duped. Digital signatures, which will identify the originator of a communication and indicate the originator's approval of the content of the communication, should help to address this issue.

There have been several policies put forward regarding how certain issues might be addressed. Whilst these have much to commend them, computing activity is globally interactive and yet these policies tend to be introspective. There appear to be policy vacuums regarding electronic commerce which include addressing authentication, encryption, time, and place of contracts, information collecting agents, aggressive marketing through electronic junk mail and subliminal advertising and universal access to all types of products and services. All these and no doubt others need to be addressed from a global perspective.

Shopping is a social activity. As well as its functional role it includes the pleasure of browsing, impulse buying, discovering new shops, casual conversation, and planned and unplanned meetings with other people. Electronic shopping is not the same. It is primarily a functional activity since most of the social activity cannot or is not supported. You cannot bump into a friend in the electronic shopping precinct. Window shopping is hampered by the primitive nature of search engines. The consumer is dealt with automatically by the computer systems of the supplier—it is a person to machine relationship and not a person-to-person relationship as is the case with traditional shopping. Whilst electronic commerce provides access at all times and provides a lot of benefit to certain groups, such as those who are house bound, it is very limited in terms of its social welfare capability.

Electronic commerce will have a profound effect on society and its organisations and upon our lives as consumers and employees. It is not clear what that impact will be. Therefore, a broad and continued social dialogue about these questions is essential to overcome the hurdles and reap the benefits.

3.7 Identity Crisis Online [2000]*

This section was first published as: Rogerson, S. & Wilford, S. (2000) ETHIcol—Identity crisis online. IMIS Journal, Vol 10 No 3, pp. 28–29. Copyright © Simon Rogerson & Sara Wilford.

At the recent Computers, Freedom and Privacy Conference 2000 in Toronto, Canada there was an interesting session on identification and authentication in the on-line world. The panel consisted of Deirdre Mulligan—panel co-ordinator, Karl Ellison from Intel, Phil Hestor of IBM, Margot Saunders a lawyer at the National Consumer Law Centre USA and David Flaherty the ex-privacy commissioner for British Colombia. This edition of ETHIcol considers some of the main issues raised during that session.

How do you prove you are who you say you are? How do you know that someone is legitimate in his or her dealings with you, and how do you get redress if things go wrong? If your identity is stolen and used fraudulently, or personal records are altered without your knowledge or permission how do you prove that it was not you? It is difficult enough to verify someone's identity in the tangible world where forgery, impersonation and credit card fraud are everyday problems related to authentication. Such problems take on a new dimension with the movement from face-to-face interaction, to the faceless interaction of cyberspace.

The world of cyberspace has many difficulties of identification and verification due to its remote and electronic nature. You can never be sure with whom you are dealing or if the goods and services you are attempting to buy even exist. This is why the "digital signature" and other authentication systems are being developed. Identity is however, important not only between individuals and organisations and from person to person, but also to promote trust in on-line companies and to verify their legitimacy.

In the UK, for example, these issues are being addressed by two initiatives which are in line with the E-communication Bill. The tScheme will help to deliver an industry-agreed code of practice governing the way digital certificates are issued and managed for the purposes of legally binding communications. TrustUK will accredit e-commerce codes and provide a "hallmark" that accredited companies may use on their websites.

There are, however, problems associated with digital signatures, even with the use of biometrics. For example, the use of a thumb print reader may only provide the equivalent of a three or four digit password, which is easily hackable by anyone with sufficient determination and know-how. The temptation is to presume that the certified key holder is the one actually holding the key. The authorising mechanisms may actually see the access as legitimate even though the authorised person may not be the one using the system.

With the sheer size and scope of the on-line environment, it is becoming clear that the use of only names is increasingly infeasible for authentication purposes. This is because it is impossible to be sure that the people you are dealing with are who you think they are. However, key-signing sessions and other methods attempting to vouch for a key holder still only identify the legitimacy of the key, and not the user.

Whilst the need for verification to promote e-commerce is relatively clear, the needs of business and governments in verifying identity must be carefully considered in the light of individual privacy and the increasing requirement that individuals reveal more and more details about their personal lives. Are we in danger of becoming so transparent to the data banks that the privacy of the individual is only to be found inside one's own mind? The unique data that will be required to verify identity will need to be carefully protected to ensure that potentially sensitive personal information does not enter into the public domain.

The use of very strong cryptography has been cited as one way that verification and authentication may be achieved. However, the need to enter several codes and to perhaps also provide a biometric identifier is unlikely to be workable in order to purchase an item from the internet due to the sign-on burden placed on the consumer, and indeed may undermine the principle of easy on-line shopping.

As consumers, we need to be assured that our credit and debit card details do not go astray, and that only those documents with our authorisation and verification will be acted upon. The idea that someone may use our identity for their own means, or that third parties may access sensitive information is of great concern to many. The use of authentication and security techniques is vital in addressing these concerns. The use of a credit or debit card in an on-line environment is relatively safe due to the legal obligation of banks to reimburse customers. However, commercial organisations do not have as much to lose as the customer, particularly with regard to sensitive personal data, so they may not have the incentive to provide strong security. This lack of bargaining power by the consumer can be considered problematic and may only be overcome by strong legal pressure and monetary penalties on the holders of data.

The problems associated with authentication are not just related to the verification of identity but also involve greater public policy issues, which include the amount and kind of data required to confirm the identity of someone. The use and access to such data is an issue of major importance due to its potential for abuse by organisations seeking to maximise profits by using the data for marketing purposes. The confirmation of individual identity or the authentication of those accessing or amending data is an emotive issue. There has to be a fair balance between upholding personal rights whilst enabling authenticated and secure access to on-line services and products.

3.8 Keeping Records Safe [2000]*

This section was first published as: Rogerson, S. & Fairweather, N.B. (2000) ETHIcol—computer ethics. IMIS Journal, Vol 10 No 5, pp. 26–27. Copyright © Simon Rogerson & N Ben Fairweather.

Healthcare computing or medical informatics is one of the fastest growing areas of information and communication technology (ICT) application. It is a multifaceted application concerned with electronic patient records, performance indicators, paramedical support, emergency service, computer aided diagnosis, clinical governance, research support, and hospital management. Its use should ideally promote and must certainly not be in conflict with the fundamental medical ethical principles of beneficence (a duty to promote good and act in the best interest of the patient and the health of society), nonmaleficence (a duty to do no harm to patients), and respect for patient autonomy (a duty to protect and foster a patient's free, uncoerced choices).

The Electronic Patient Record (EPR) is indicative of the advances in medical informatics and allows providers, patients and payers to interact more efficiently and in life-enhancing ways. It offers new methods of storing, manipulating and communicating medical information of all kinds, including text, images, sound, video and tactile senses, which are more powerful and flexible than paper-based systems. The policy of governments appears to favour a national healthcare infrastructure with a longitudinal patient record covering a patient's complete medical history from the cradle to the grave. Such developments raise several ethical issues.

EPRs can facilitate the doctor-patient relationship through use of computerised notes which the doctor and patient share and contribute to. However, EPRs can harm the relationship and undermine trust. For example, in the US there are medical data clearinghouses that sell medical patient data to insurance companies, police departments, employers, drug companies, and so on. Consequently, patients are becoming reluctant to tell their doctors everything about their medical symptoms and the causes of them. This is damaging the doctor-patient relationship, which depends heavily upon confidentiality and most importantly threatens to damage quality of care. There is clearly a tension and trade-off between the need-to-know and the right to confidentiality which must be addressed. This is an issue which has been exacerbated by ICT. Violations of medical confidentiality may appear to be easier because of the efficiency of computerised systems. The damage to the patient whose confidentiality is violated may be proportionately greater because of the amount of data held within the EPR.

Electronic access to medical information requires careful scrutiny. A patient's right to informed consent should prevail. A patient should have control over her/his data preventing casual distribution that might be harmful. A patient should also have the right not to be informed of medical facts, for example, genetic data which might affect self-esteem and the way in which one lives one's life. The ability to invoke only partial access is important. For example, a woman who had had an abortion might visit a surgeon for an unrelated ailment. She should have the right to decide whether or not to allow that physician access to data other than what could be termed as her general health status. Electronic access to patient data can be beneficial. For example, the administering of prescriptions via electronic transactions from doctor to pharmacist to patient with the aid of digital signatures for authorisation could improve service significantly.

The movement of EPRs over the internet, intranets and extranets raises concern, for the further from the original source the greater the risk of inaccuracy, falsification, duplication, manipulation, and unauthorised distribution. There is little effective control over data use over computer networks. Ethical practices are not well defined for the vast array of disclosures to secondary users, such as managed care evaluators, insurance companies, and so on. For example, categorising and profiling patients may engender discriminatory and or exclusionary effects. Data banks of health maintenance organisations, and drug companies are gathering information and storing it in computerised form. By linking their computers these organisations can trade information across computer networks. Information has thus become a tradable commodity. The legitimacy of this medical information trading needs to be established and assurance sought that data is completely anonymised.

A morally defensible approach to EPRs should be adopted and could be based on the following:

■ A patient's right to informed consent should be dominant: thus education about these issues should be available to patients, along with information about the existence of all databases with medically relevant information about the patient.

- A patient should normally have effective control over his/her data and the ability to prevent any casual distribution that might be harmful, ensuring EPRs maintain nonmaleficence. There need to be safeguards to ensure that declining to give consent to access records does not harm the patient unduly.
- Patients should have the ability to allow selective access to their records.
- The EPRs of the dead should be treated with the same consideration as those of the living.
- In contemporary industrialised societies, legislation should clearly define the appropriate scope for EPRs, and ownership of patient data. It should clarify what principles should govern legitimate access to and use of personal health and medical data and information, and patients' rights with respect to their own medical information. There should be prohibitions on certain sorts of uses of data. Mechanisms for the adequate enforcement of applicable laws and oversight of use and access must be in place.
- Healthcare providers and funders and other potential recipients of medical data should understand the range of impacts of all sorts of medical data sharing, including on the requirement for openness in the doctor-patient relationship (both if patients are to be appropriately treated and if accurate data is to be collected).

EPRs are indicative of a society that is increasingly dependent upon ICT. The impact of this morally sensitive application of ICT cannot and should not be ignored. Policy makers, key decision takers and developers involved in the creation, use and promotion of EPRs are urged to consider these suggestions.

3.9 Health Warning [2001]*

This section was first published as: Rogerson, S. & Fairweather, N.B. (2001) ETHIcol—Health Warning. IMIS Journal, Vol 11 No 5, pp. 26–28. Copyright © Simon Rogerson & N Ben Fairweather.

Health is of concern to all of us. There is increasing use of the worldwide web to seek information about health-related issues. Indeed, we are currently in a situation with massive amounts of healthcare information on the web. Healthcare professionals now have practical access to information that often was inaccessible to them in the past. Many of the serious issues with healthcare websites come when we think beyond their use by professionals. Members of the general public are much less able than professionals to evaluate the sites that they find, but even professionals may not always evaluate sufficiently.

This article focuses on the use of healthcare websites by the general public. A recent survey in the US by Harris Interactive found that 47% of all adults use the internet and of these 75% or 100 million go online to seek healthcare information. They do this on average three times a month.

A Boston Consumer Group survey found that this online information has a real impact on the way people manage their overall care and comply with prescribed treatments. Indeed, the Ipos-NPD Online Health Report states a high level of satisfaction with the quality of healthcare websites particularly on specific conditions and diseases. According to the report the greatest trust is placed in thebreastcancersite.com, americanheart.org, WebMD.com, and DrKoop.com. Interestingly, those sites with the lowest trust were dieting, health insurance/HMO/managed care and brand-name prescription sites.

Is this level of consumer satisfaction and trust misguided? The study, published in a recent edition of the Journal of the American Medical Association, found that numerous healthcare sites aimed at residents in the US left out key information, offered out-of-date information, or offered contradictory information. Some sites blurred the distinction between advertising and medical advice. As part of the survey a team of physicians examined the content of 25 healthcare sites. They found around 50% of the information on English-language sites and 20% on Spanish-language sites included advertising that was not clearly labelled as such.

It is clear that members of the public are less able to appreciate the subtleties of information presented. Much is written in specialist language, but even without specialist language, the reading level would be beyond most of the public. This could mean that patients might discover unbalanced information or misunderstand the subtleties of good information resulting in unnecessary worries and extra demands on family doctors. Alternatively, it could mean that patients worried about their condition might get inappropriate reassurance from web-based information and therefore do not seek professional advice when they need to.

One way to better ensure patients get the correct message is to use one-to-one advice using a web or email interface. However, this opens up fresh challenges in terms of security of data transmission and storage. Some of these are serious technical challenges whilst others are organisational. Even if the security challenges are met, patients may only be prepared to reveal relevant facts if they have confidence that they can control to whom the information is subsequently released and they may wish to instigate selective release of such information. Clearly there is a need to verify and evaluate the status of those giving the advice. While some may see web-based advice as a chance to obtain anonymous advice, the providers of such advice will probably want to guard against timewasters and there are few ways of doing this without users divulging their personal identity. This presents problems because it is easier to impersonate someone electronically than in a face-to-face consultation with, for example, a family doctor.

Issues of confidentiality and security are not confined to cases where the patient explicitly gives information to a site since, for example, standard logs of web access can identify which pages were used and the host name of the computer that accessed them. In many cases this is not a problem, but some organisations routinely give workstations the name of the person who uses them, and it is therefore often possible to surmise something of, for example, the health status of the person searching healthcare sites. Thus, if Joe Soap uses his office computer, soap.dept.company.co.uk to access a lot of information about testicular cancer, and the prognosis for patients with such a diagnosis, life insurers to whom Joe is applying for extension of cover might be very interested in such information. They would be unaware of the fact that Joe accessed the information for a colleague. Similarly, such information may be used by those wishing to advertise or sell treatments and therapies. Access to such information and the resulting actions might be both inappropriate to the patient and illegal.

There is statistical evidence to suggest that advertising may also cause websites to be distorted. Distortion may range from sites that are designed to promote a particular product to those which aim for impartiality, but (perhaps unconsciously) play down or omit criticism of products advertised. Worse, sites may seek to obscure their relationship with those that pay for them thus resulting in manipulation of those seeking information. Such sites may either be commercial or seeking to promote a position on, for example, abortion, assisted suicide or voluntary euthanasia.

Validating seals may help patients know which sites to trust, but to be meaningful they need to only be awarded after detailed examination of the site. Schemes such as TNO Health Trust are being launched but it is unclear whether the cost can be sustained and how many of the thousands of sites can be vetted in practice. Moreover, medical knowledge changes and this presents currency and integrity problems both for providers of information and for those validating such information

There is one final note of caution. Technological limits inherent in the architecture of the web or caused by viruses and attacks mean that as things are now, we cannot rely on web services in life-critical situations such as those related to health. Whilst it may be possible to provide access to guidance from remote specialists the problem is that an increasing use of such facilities may mean that we come to rely on them and trust them unquestioningly which at best is unwise and at worse could be catastrophic.

The wide variety of problems associated with web-based health information means that the cost savings that some see as being possible from the use of the web in healthcare may well be illusory. Such applications of new forms of technology require new approaches particularly in the initial stages of development and implementation in order to ascertain the true costs and benefits.

3.10 Facing Up to the Issues [2002]*

This section was first published as: Rogerson, S. (2002) ETHIcol—Facing up to issues. IMIS Journal, Vol 12 No 1, pp. 27–28. Copyright © Simon Rogerson.

Surveillance has become part of everyday life. The compelling arguments for its introduction into public spaces as means of ensuring public safety have led to CCTV cameras perched high above every street corner within towns and cities. The next generation of surveillance, smart CCTV is now being introduced under the guise of routine maintenance and upgrade. In June 2001 100,000 people attending the Super Bowl in Florida had their faces scanned in search of wanted criminals.

Smart CCTV is the combination of video surveillance with facial recognition technology and is used for face finding, face recognition and face tracking. Advances in facial recognition technology, using faceprints based on facial geometry, has been significant. According to Visionics Corporation, it is now possible to search one million faceprints per second with an error rate of less than one percent. Contextual factors such as lighting, expression and profile have no real impact if the image is of good quality.

Philip Brey of the University of Twente in the Netherlands discussed the ethical issues of smart CCTV in public spaces in a paper he presented at the CEPE conference held at Lancaster University in December 2001.

He reports that in the UK smart CCTV has been used in the London Borough of Newham since 1998. At the time the following was reported in Computer Weekly (13 October 1998):

> The London Borough of Newham aims to cut crime by 10% within six months by using face recognition software to pinpoint criminals on CCTV cameras. The pilot, to be jointly launched tomorrow with the Metropolitan Police, is the first local authority implementation of the technology. Newham has already fielded enquiries from 20 councils and eight police forces, and Bob Lack, Newham's emergency services manager, predicted that many of the 250 councils in the CCTV user group would soon adopt the technology. But the project has drawn criticism from the Data Protection Registrar, which has voiced concern over the implications for privacy. The Registrar is seeking a meeting with the Metropolitan Police. "People are being compared to convicted felons—there are clear civil liberties implications," said Jonathan Bamford, Assistant Data Protection Registrar.

In 2001 the Newham system was linked to a central control room operated by the London Metropolitan Police Force. In April 2001 the existing CCTV system in Birmingham city centre was upgraded to smart CCTV. People are routinely scanned by both systems and have their faces checked against the police databases.

Brey explains that the debate about smart CCTV has primarily centred around the security benefits of the technology versus the threat to privacy and freedom. He points out that it is vital to understand thoroughly the trade-offs that have to be made between security and civil liberties when considering how and where to use smart CCTV. "A better understanding is needed of both the importance of civil liberties and the importance of security, of power and reliability of the technology, and of its potential uses and abuses." Brey identifies three ethically charged problems associated with smart CCTV.

1. Error—This is where incorrect matches result in innocent people being subjected to investigation and harassment by the police. Brey suggests that "this problem does not in itself present a strong argument against smart CCTV. It only suggests that every care should be taken to minimise error, to minimise the inconvenience to mistakenly identified citizens, and to evaluate whether the trade-offs that are made are still reasonable."

2. Function Creep—This is where the purpose for using smart CCTV may be easily extended from identifying criminals and missing persons to include other purposes. This could occur by widening the faceprint database to include, for example digitised images from driving licences. It could occur by undertaking analysis of crowd behaviour and membership or tracking individuals over a period of time to ascertain movement and interaction behaviour. It could occur by the system being used by new types of users for example international law enforcement agencies or local town councils. Individual operators could use, albeit unauthorised, the system for their own purposes. Brey argues that unacceptable function creep is a common problem with all new technologies and cannot be wholly avoided by regulation. He suggests that there is "an obligation on developers and users of technology, therefore, to anticipate function creep and to take steps to prevent undesirable forms of function creep from occurring."

3. Privacy—Intrusions on privacy through observation of individual behaviour is a common problem with all surveillance. Smart CCTV suffers from an additional problem. The face is a "highly personal aspect of one's body" and smart CCTV captures this in digital form as a faceprint. Brey points out that in this context the face is nothing more than an information structure. He explains that "the unique features of one's face, by which others recognise you and which help to define your uniqueness, can be encoded into a computer file of only 88 bytes. This functional reduction of body parts to information structures is one that many people find dehumanising." He points out that "the faceprint that uniquely characterises your face is not "yours" but "theirs': it is not owned by you or even if it were, it would not be understood by you because you do not understand the technology." Thus, smart CCTV poses two privacy problems.

Whilst smart CCTV does provide social benefit, it does require a reduction in civil liberty. People do have justifiable privacy expectations even when they are in public space. Such expectation can be violated by smart CCTV. Furthermore, faceprints are part of our electronic persona and as such must be treated as part of us. It is unlikely that developers or those applying this technology have considered such issues in depth. The call must be for an effective social audit process when such technological applications are being considered. This process should capture and take into account the opinion of those directly and indirectly affected.

3.11 Clear and Present Danger [2002]*

This section was first published as: Rogerson, S. (2002) ETHIcol—Clear and Present Danger. IMIS Journal, Vol 12 No 2, pp. 27–28. Copyright © Simon Rogerson.

On Friday 2 June 1994 a Chinook helicopter ZD576 crashed on the Mull of Kintyre killing 29 people. Pilot error was suggested as the cause. *Computer Weekly* has been campaigning for this decision to be overturned as there was evidence that the crash could be a result of systems failures on the helicopter. On Wednesday 6 February 2002 the House of Lords committee report found that there is doubt about the cause of the crash because of the possibility of a technical malfunction.

Extracts from various press articles and reports illustrate the broader issues related to the development of safety critical software development.

3.11.1 Victory! Lords Confirm CW Stand (7 February 2002)

In 1994, 29 people, including some of the UK's top anti-terrorism experts, were killed when Chinook helicopter ZD576 crashed on the Mull of Kintyre. In the absence of any clear explanation for the accident, the crash was blamed on the pilots of the helicopter who were killed in the crash.

Since 1997 *Computer Weekly* has published details of the compelling evidence that has emerged that the crash could, in part or in full, be a result of systems failures on board the helicopter.

In 1999 *Computer Weekly* also published *RAF Justice* which made clear how the Royal Air Force had carried out a cover-up and blamed the pilots unfairly.

Yesterday the House of Lords endorsed *Computer Weekly's* findings, putting pressure on the government to overturn the Ministry of Defence's verdict of gross negligence against the pilots of Chinook ZD576. While it is imperative that justice is done, it is also vital that the Chinook saga is recognised as an example of just how the severe the consequences of poor IT project management can be.

3.11.2 Software Flaw Could Have Caused Chinook Crash, Tony Collins (7 February 2002)

The Lords committee report, which names *Computer Weekly* as having provided information to Parliament, found that there is doubt about the cause of the crash because of the possibility of a technical malfunction, such as a jam of the pilot's controls or a sudden engine surge, caused by the Chinook's safety-critical full authority digital engine control (Fadec) system.

The Lords committee's verdict: "We have considered the justification for the air marshals" finding of negligence against the pilots of ZD576 against the applicable standard of proof, which required "absolutely no doubt whatsoever." In the light of all evidence before us and having regard to that standard, we unanimously conclude that the reviewing officers were not justified in finding that negligence on the part of the pilots caused the aircraft to crash."

Software problems: "It is clear that at the time of the crash there were still unresolved problems in relation to the Fadec system of Chinook MK2s."

Boeing's simulation: "We consider that Boeing's conclusions cannot be relied upon as accurate." (Boeing's simulation was crucial to the 1995 enquiry's conclusion that the pilots were in control.)

3.11.3 Lessons to be Learned from Chinook Tragedy, Tony Collins (7 February 2002)

Be on top of the project, or you will be almost entirely in the hands of the supplier if there is a disaster.

When a user company passes day-to-day control of a project to its IT supplier and then suffers a serious software-related problem or even a major software-related fatal accident, there may be no sure-fire way of establishing what has gone wrong or why.

This is because manufacturers cannot be expected to indict themselves after a major incident by identifying defects in their product or management of a project.

The problem with software is that only the manufacturer may understand it well enough to know what has caused or contributed to a crash; and even if the supplier wants to tell the whole truth, will its lawyers let it?

One solution may be to employ independent experts to scrutinise the manufacturer through-out a project—it may be too late to employ them after a disaster.

3.11.4 Clear and Present Danger: Why CW Refused to Give Up on Chinook, Karl Schneider, Editor, Computer Weekly (7 February 2002)

The most worrying single facet of the crash is that the Ministry of Defence and the RAF assumed that a lack of evidence of malfunction points to operator error being the cause.

What if a software problem caused the accidental firing of a missile that destroyed a town? What if a design error in a software-controlled train set off a complex sequence of events that caused a fatal crash? What if dozens of people were killed in a fire in a tunnel because software-controlled sprinklers failed?

The chances are that if software caused any of these accidents, we would never know. This is because when software fails, or it contains coding or design flaws, and these defects cause a major accident, there will be no signs of any software-related deficiency in the wreckage.

And only the manufacturer will understand its system well enough to identify any flaws in its design, coding, or testing. Yet no commercial manufacturer can be expected to implicate itself in a major software-related disaster. So, if software kills large numbers of people it is highly likely that the cause of the accident will never be known.

This is especially likely to be the case if the software has failed in no obvious way, such as when a coding error has set off a chain of complex events that cannot be replicated after a disaster.

But after a major accident, convention dictates that someone must be blamed. Step forward the vulnerable equipment operators: the pilots, keyboard clerks, train drivers or anyone who cannot prove their innocence.

This is particularly so because the manufacturer, in proving its equipment was not at fault, may have millions of pounds at its disposal. It will also have the goodwill of the customer, which bought the highly specialised equipment and relies on the manufacturer's support for its maintenance.

In contrast, individuals—the system operators—may have minimal resources: no access to the manufacturer's commercially sensitive information, none of the manufacturer's knowledge of how the systems work, and little money for expert reports and advice.

Therefore, the weakest link after a major fatal accident will always be the operators—particularly if they are dead.

That is why the loss of Chinook ZD576 is so much more than a helicopter crash. To accept the verdict against the pilots is to accept that it is reasonable to blame the operators if the cause of a disaster is not known.

If we accept this dangerous principle we may as well say to manufacturers of safety-critical software, "We recognise you will try to do a good job, but if you create poorly designed, haphazardly developed and inadequately tested safety-critical computer systems that kill people, we acknowledge that you will never be held to account."

3.11.5 Boeing Simulation Ignored Chinook's Fadec System, Tony Collins (15 November 2001)

A Boeing simulation of the last moments of flight took no account of the helicopter's new Full Authority Digital Engine Control (Fadec) computer system.

In extreme cases, the Fadec could cause the Chinook's jet engines to surge suddenly, making it difficult to control the helicopter.

3.11.6 RAF Justice—a 140-Page Report on a Cover-Up of the Chinook's Software Problems Published by Computer Weekly in 1999

For reasons of safety, each FADEC was to have two "lanes" which performed similar functions. The main or primary lane was to be a computer system. The back-up, or as it called "reversionary" lane, was to be based on more conventional analogue technology.

But as time went on the project became more technologically ambitious and, without any opposition from the Ministry of Defence, the manufacturers went ahead with a system which was digital in primary and back-up mode. Unusually in a FADEC system, there was no mechanical backup.

3.11.7 A Warning to IS Professionals

The integrative nature of today's software means the lines are increasingly blurred between a product's components. This case illustrates the need to accept that those developing software have a joint obligation with other professionals in delivering safe, usable and trustworthy products. Project and post implementation management procedures must ensure products are effectively tested and monitored. Abnormal events must be thoroughly investigated and corrective action take. The fundamental principle must be to safeguard the public.

3.12 At Cross Purposes [2002]*

This section was first published as: Rogerson, S. & Fairweather, N.B. (2002) ETHIcol—At Cross Purposes. IMIS Journal, Vol 12 No 3, pp. 26–27. Copyright © Simon Rogerson & N Ben Fairweather.

The UK Government is pushing for all service delivery to be e-enabled by 2005 and as part of this process is looking at electronic voting. After all, a vote is, essentially, a piece of information, and the public are already used to voting by telephone and the internet for television shows. Against this background, ministers such as Robin Cook have been keen to point out that voting by marking a cross in pencil on a piece of paper in a plywood polling booth seems to come from another era. The recent local elections in England and Wales have included several e-voting pilots as part of the Governments move to full implementation.

Electronic voting offers real advantages including the possibility of voting from more convenient locations and the chance for disabled voters to vote on equal terms with others.

However, voting is one activity where equal access is of vital importance: Robert Mugabe sought to influence the result of the Zimbabwe presidential election by making it easier to vote for those who he thought would vote for him, and more difficult for those he thought would vote against him. Unlike most public services, when it comes to voting, it is not good enough to make voting much easier for some people without making it easier for all. Voting systems cannot rely on a technology that some have easy access to, but a significant proportion is struggling to use. This alone seems to suggest that simply making it possible to vote using home computers will be unfair: less than about 40% of voters in the UK have access to PCs in the home.

E-voting also raises tough issues of security and secrecy. Virtually no other transaction with Government requires the same degree of secrecy, even from other family members. History has shown that if secrecy is not maintained, there will be attempts to buy votes, to coerce voters and to use undue influence in other ways. It is no good, therefore, if problems are encountered, to go back to receipts as in other electronic transactions: receipts are just what would be needed to buy and sell votes and make sure coercion has worked.

If voting is allowed away from supervised polling stations, how do we know that there is not someone else in the room coercing the voter? If we get to the stage where systems are routinely supplied with cameras, then it might be possible for there to be checks that the voter is not being coerced, but this in turn will cause worries about privacy and could be very labour intensive.

There is a clear requirement for the election authorities to know that each person who is submitting a vote is genuinely entitled to submit a vote and has not already done so. Yet at the same time, the requirement for secrecy means that how the voter has voted should not be revealed to those in authority. This problem does not seem to be logically insoluble: issuing every voter in an election an identity code for that election, but which cannot be improperly linked back to the identity of the individual is part of the solution.

A fundamental problem for electronic voting or electronic counting of votes, is how do we know that we can trust the systems to give us a "result" that accurately reflects how people have cast their votes. This is not just a case of how we avoid farces like Florida. The question of knowing whether we can trust systems goes deeper.

If you use ICT to cast a vote, how do you know that when you tell it "vote for candidate X" it really does so? It might tell you that it has, but how do you know it is not telling lies? Part (but not all) of the problem here is with viruses. If you are using a multi-purpose computer with a mainstream operating system, you are wide open to virus attacks. Sure you might have "virus protection," but virtually all "virus protection" systems only protect you against viruses that they already know about, and there is no guarantee that it will know about a virus that changes votes as you cast them.

Assuming that all votes that are sent do genuinely reflect the intentions of the voters, how do we know they have been counted accurately? We could ask the computer to count them again and be amazed when it comes to exactly the same result as last time (unless we are using the punch cards they used in Florida). But that does not tell us that it has counted them correctly either time. Part of the solution here is for the software that does the counting to be open to inspection by the political parties (or experts they nominate), but that does not tell us that it has been compiled in the way that we expect. So the compiler needs to be open to inspection, and the computer that the compiler runs on, and so on, and so forth. Transparency is therefore key.

Even if we are confident the votes in the database are correctly counted we also need to know that they were recorded correctly in the first place, and the database has not been improperly modified. It is far from easy to ensure this in the face of people who might be determined to rig an election.

Although it is not perfect, marking your vote in pencil on a piece of paper in a plywood polling booth still has much to commend it. The challenge for those thinking about e-voting is to come up with something that improves on that, but without throwing away the essential advantages that pencil and paper have. Politicians through to developers must work together in a socially responsible way to ensure electronic voting promotes the democratic process. They must have the courage and conviction to reject the technological solution if this is not the case.

3.13 Matching Cases [2002]*

This section was first published as: Rogerson, S. & Wu, X.J. (2002) ETHIcol—Matching Cases. IMIS Journal, Vol 12 No 5, pp. 27–28. Copyright © Simon Rogerson & Xiaojian Wu.

In February 1997 this column focused on the potential ethical problems associated with deploying data matching techniques by both public and private organisations (see Section 5.1). In light of the advantages in information sharing through expanding communication networks amongst more and more organisations, two issues were highlighted about the practice of data matching that are of moral significance to the public. The first was the accuracy of data being handled and the second was the security of networks. It was suggested that there must be a clear understanding of who has access to such information and how such information will be used. It was stressed that data matching must be used in a balanced way that is sensitive to the needs of organisations and society as a whole as well as to the rights of individuals.

Not surprisingly in the so-called Information Society or global village, technologies, which enable data matching and information sharing, have been introduced over the last five years into various social environments at a level not seen before. But to what extent does this kind of activity accommodate or address the two ethical concerns already mentioned? Consider the following Chinese case study.

China has, for the last half a century, a unique civil administrative practice called the household registration system, otherwise known as *Hukou*. Under this system, all Chinese citizens are mandated to register with the police their personal information including their residential address, religion and employment details, and, within a stated period, report any permanent or temporary changes. Since the mid-1980s, every Chinese citizen above age 16 has been issued with a Resident's Identification Card (RID), which holds basic information about the individual and his or her permanent address. Citizens are legally required to carry and produce the RID when and where necessary. Police stations within districts in China are responsible for operating this system. Before the 1980s, Hukou administration in a police station was manual and typically involved a number of police officers specifically tasked to do the registration work and maintain the manual filing system.

In the early 1980s, the Chinese police began to take advantage of the latest information and communication technologies in discharging their work and services. Computerisation of the household registration record has brought a significant transformation to policing in China today and·has had a significant impact on the life of most Chinese urban residents. Initially, most of the household registration records were stored in individual computers located in diverse areas. Hukou administration remained essentially the same as the police merely automated the registration process and replaced manual files with computerised files.

Since the mid-1990s, with the development of database and networking technologies, the Ministry of Public Security, which is in charge of the Chinese national police, became aware of the potential benefits of connecting the diversely located computers into a national police information network. Consequently, it officially launched in 2001 the "Golden Shield" project, which aims to construct a multi-tiered police information web enabling communication and sharing of information amongst the Chinese national police, and improving working efficiency. The household registration information database is the backbone of this project with a variety of police operational systems such as crime prevention and detection utilising this database.

In 2001 the Ministry of Public Security was instructed to tackle the increasing number of offenders and primary crime suspects who were moving away from their original places of residence. This group continued to offend as they moved from one location to another. As part of the clamp down police at different levels and locations within the multi-tiered national police information network were advised to upload the information about known criminals or crime suspects. This information usually consisted of the crime committed and the person's RID information. Police throughout China were then able to access this information across the network and then search their own computer records of household registration for possible and probable matches.

As part of this campaign, the police were also encouraged to stop all people on the street they deemed to be "suspicious" and to check them against the centrally updated wanted-criminals database. The police took action once they identified a match between information on a person's RID and information on the database. This action was normally a detention and investigation of the person. This campaign lasted for two months and resulted in hundreds of thousands of arrests and charges throughout the country.

In terms of solving committed crimes this approach of information sharing and matching proved very successful. However, there were concerns about the integrity of the information being uploaded and downloaded. One particular incident happened in a city in the North Eastern Liaoning province, where one local police officer, for personal reasons, had uploaded onto the system the crime and personal information of three local people who had already been convicted and served their prison terms. This led to the three people being re-arrested and charged in a faraway city where they had just started their new normal life. Such kind of incidents or malfunctions of the system were not isolated during the two-month campaign. This has caused grave concerns amongst the general public, particularly the ever-increasing floating populations in many big cities in China because they are the primary target for such data matching during this type of policing campaign.

There is much to be learnt from this Chinese example. Understandably, this case study exhibits many social characteristics of the Chinese society, where the regime and culture are quite different from that in western countries. Nonetheless, the attraction of increased processing capabilities of new computing technologies and the appetite for more information by both public and private organisations is unlikely to be limited by national culture or political systems. Our fundamental human values could be compromised, and indeed will be compromised, if we do not treat with serious moral considerations the application of computing systems to data profiling, data matching, and information sharing.

The ethical concerns raised in the February 1997 article still remain well founded. It is important to remain vigilant and to call for appropriate legislative frameworks, codes and operational procedures to relieve us of the worries associated with this type of computing application.

3.14 Going, Going . . . Gone! [2003]*

*This section was first published as: Rogerson, S. (2003) ETHIcol—Going, going . . . gone! IMIS Journal, Vol 13 No 1, pp. 26–27. Copyright © Simon Rogerson.

Internet auctions seem to be one of the few electronic commerce facilities to capture the imagination of consumers around the world. The potential is huge given the online world population stands at 606 million (NUA statistics for September 2002). In October 2002 Forrester Research predicted that by 2007 internet auction sales will account for 25% of total online retail sales in the US. This would be an increase from $13 billion in 2002 to $54 billion in 2007.

However, there are concerns by many about the trustworthiness of internet auctions. This is fuelled by continuing reports of shocking incidents concerning either the types of goods being sold or the actions of those involved in auctions. Here are just a few examples of reported incidents.

In a sophisticated con, fraudsters are copying information about real cars for sale on sites such as AutoTrader.com and CACars.com and are listing them for sale on eBay Motors. Buyers purchase the cars from these con men, wire-transfer money to a seemingly legitimate escrow service and await delivery of their car. When the car never arrives, they contact the real owner of the car, and discover the legitimate seller knows nothing about the transaction. Then the buyers realize their mistake: the escrow service they've sent the money to is a scam, and the "seller" they dealt with never had possession of the vehicle.

(Online Escrow Fraud Hits eBay Members, Ina & David Steiner, Auctionbytes-NewsFlash, Number 421 — 25 October 2002 — ISSN 1539–5065)

Shill bidding involves a seller directly or indirectly placing a bid on his or her own item in order to encourage legitimate bidders to bid higher. . . . It's possible that eBay doesn't see all shilling as a terrible thing, on the theory that shill bidding early in an auction can be regarded as harmless "pump priming." Shilling used to be within eBay's rules, in fact, and even today, according to Pursglove, such low-key shilling is unlikely to inspire any official action. Unless, apparently, the pump priming works a little too well: Last May, a California lawyer with at least five eBay identities was booted off the site for entering a $4,500 bid on the second of 10 days of bidding on a painting rumored to be by well-known California artist Richard Diebenkorn, kicking off a bidding frenzy that saw the price climb to $135,805. EBay later voided the sale.

(Sleaze Bay, David H. Freedman, Forbes ASAP, 27 November 2000)

A California man who collected $36,000 from bidders over eBay's internet auction site then failed to deliver promised goods was sentenced to 14 months in prison, federal authorities said. . . . Guest has admitted to posting several listings on eBay's Web site that fraudulently offered Sony digital cameras, IBM laptop computers, and other items.

(Bloomberg News, Special to CNET News.com, 2 November 1999, 2.10pm)

Only yesterday, a seller on Amazon's auction site was accused of bilking customers of $28,000, collecting payments from buyers but not shipping the goods. Amazon, which is investigating the case, says if it is true, it would be one of the worst cases of fraud on its auction site. And last month, eBay closed an auction of 500 pounds of marijuana battled a spate of auctions involving illegal sales of human organs-and even an unborn baby.

(Net crime poses challenge to authorities, Troy Wolverton and Greg Sandoval, Staff Writers, CNET News.com, 12 October 1999, 2.25pm)

Three men attempted to auction 500 pounds of marijuana on eBay Tuesday night until company officials pulled the plug-almost a day after the sale began. The sale began roughly 8:40 p.m. PT on Tuesday night and ceased sometime 21 hours later, according to a screenshot on AuctionWatch.com, which first reported the story.

(eBay auction goes up in smoke, Greg Sandoval, Staff Writer, 23 September 1999, 6.30pm)

There is a catalogue of problems which eats away at the trustworthiness of online auctions. The types of goods being offered are open to question. Are they illegal items in certain parts of the world or are they items which might cause offence? As seen from the illustrations online auctions present rich pickings for fraudsters. Are the goods on offer real or hoax? Even the actions of those operating online auctions sites are open to question. The tracking of buyers and sellers using email monitoring and profiling techniques has been challenged by many as being unacceptable practice. The linking of suppliers of "independent" advice on goods for sale is seen as potentially influencing the relationship between bidder and seller.

Public pressure and legal challenges have led to auction sites evolving policy to address such problems. For example, eBay.co.uk lists the following policies on its website:

- **User Agreement** Agreement spelling out your relationship with eBay.
- **Listing policies** Guidelines for listing items for sale.
- **Is my item allowed on eBay?** Prohibited, questionable, and infringing items.
- **Privacy policy** Your privacy rights.
- **Non-paying bidder policy** When bidders don't pay.
- **System Outage policy** Compensation policy for sellers when outages occur.
- **Board usage policy** Guidelines for users who post on eBay boards.
- **eBay employee trading policy** Rules for eBay employees who trade on eBay.
- **Question about eBay trademarks?** Email your trademark or copyright question here.
- **Feedback Removal Policy** eBay's rules on removing feedback
- **Distance Selling Regulations** Learn about the Consumer Protection (Distance Selling) Regulations 2000.
- **VAT** VAT and your eBay Sales

However, sometimes public pressure can take on an unacceptable face. Ina Steiner recently reported on a new trend (eBay Auction Fraud Spawns Vigilantism Trend, *Ina Steiner*, Auctionbytes-NewsFlash, Number 411 — 12 October 2002 — ISSN 1539–5065). She explained,

A new trend has emerged in online auctions: vigilantism. People are banding together to report auction fraud and are coordinating their efforts in contacting law enforcement officials.

Victims feel a great frustration in trying to get answers from auction sites and law enforcement officials. It is, for the most part, a one-way dialog as evidence is collected.

Do victims who band together actually help police efforts? Detective Burns of the New York Police Department Public Information Office said if victims are feeding law enforcement officials helpful information, it's "all well and good." But victims should not try to take actions on their own. According to Burns, victims should always contact their local police department when they think they have been defrauded on an online auction site.

Victims commonly feel extreme frustration that eBay does not have a way to warn other members about bad sellers other than the limited feedback method, where members have a mere 80 characters to sum up their complaints. New sites are popping up where victims can complain about trading partners, make their case, and vent their frustrations.

AuctionBlackList.com allows users to add auction fraudsters to a database. Buyers can search the database to see if sellers have a "record." Another site, eBayersThatSuck. com, encourages people to detail their bad experiences with auction trading partners.

Vigilantes are themselves beyond the law. The ability to right a false accusation appears weak in both websites mentioned by Steiner. The potential to victimise a seller seems high and such accusations have a habit of becoming the truth in the online world. Creating a trustworthy online auction environment requires responsible action and not the knee jerk reaction of the vigilante. We need to address the intrinsic nature of the internet which Liao and Hwang suggest comprises open connectivity, opacity, lack of a trustworthy payment instrument, and weak binding of identities. Such characteristics conspire against trustworthiness of online auction sites. Liao and Hwang (2001) propose that online auctions should be based on verifiable fairness which comprises the following six properties:

- Property 1. *Privacy before bidding*: a bidder cannot know the bids of others before bidding.
- Property 2. *Deadline enforcement*: the submission deadline is strictly regulated so that no one can bid when bid submission ends.
- Property 3. *Bid integrity*: during transmission and processing of bids, no bid will be extracted or tampered.
- Property 4. *Validity*: the winner bids highest among all correct bids.
- Property 5. *Non-repudiation*: the winner cannot repudiate his bid.
- Property 6. *Privacy of losing bids*: no one, including the auctioneer, can learn the content of losing bids.

Such properties of fairness seem laudable in creating an acceptable online auction environment. It is this sort of approach that will turn the community values listed subsequently that eBay propounds into reality.

eBay is a community where we encourage open and honest communication between all of our members. We believe in the following five basic values.

1. We believe people are basically good.
2. We believe everyone has something to contribute.
3. We believe that an honest, open environment can bring out the best in people.
4. We recognise and respect everyone as a unique individual.
5. We encourage you to treat others the way that you want to be treated.

eBay is committed to these values. And we believe that our community members should also honour these values—whether buying, selling, or chatting. We hope these community values will help you better understand the eBay community.

3.15 Playing Your Cards Right [2003]*

This section was first published as: Rogerson, S. & Fairweather, N.B. (2003) ETHIcol—Playing Your Cards Right. IMIS Journal, *Vol 13 No 2, pp. 27–28. Copyright © Simon Rogerson & N Ben Fairweather.*

Seven years since the UK government of the day proposed to introduce a multifaceted smart identity card, a new but somewhat similar proposal has been put forward by the current government under the guise of entitlement cards. The concerns raised in 1995 remain. Technological advances have tended to increase the criticality of such concerns as well as introduce new ones.

It is disingenuous for the Government to call the cards "Entitlement Cards." According to section 1.3 of the consultation document "A card scheme would entail: establishing a secure database which could potentially hold core personal information about everyone," which constitutes an identity database, and "issuing . . . cards to everyone on the central database," yet rather than including data on entitlements directly, the scheme would involve "linking the core personal information to other databases which held service entitlement information." It is clear from this that the cards are more closely attached to the identity database than entitlements. Similarly, if the scheme was one primarily about entitlement, it would be expected that it would be brought forward by a department concerned with entitlements, rather than the Home Office, which is much more centrally concerned with issues of policing, security, law and order.

The use of the term "entitlement card" appears to be little more than thin camouflage, and as such constitutes an underhand way to introduce a fundamental change to civil rights in this country. This is likely to increase suspicion of any card scheme.

The consultation document explains the Government's understanding of voluntary an universal entitlement card schemes:

> A voluntary entitlement card scheme would be one where: it was entirely at the discretion of the individual whether they registered with the scheme and obtained a card; it would be the individual's choice whether or not to use a card to access particular services i.e. there would always be a way to gain entitlement to a particular service without a card.

It is claimed that a "drawback of a voluntary scheme could be that those people who could most benefit from having a simple, straightforward way to assert their rights and entitlements might be among the least likely to apply for a card." This argument appears to fly in the face of theories of rational choice and is illogical. If people choose not to apply for a card, by far the most logical explanation of their action is that they judge that the balance of costs and benefits to them of having a card are such that they would not benefit from having a card. Any judgement to the contrary by the state amounts to paternalism of a sort that is at odds with a free society. In the absence of *clear* evidence that those who do not apply for a card have misjudged the personal risks and other costs of having a card, or have underestimated the benefits, this sort of paternalism cannot possibly be justified.

A voluntary entitlement card is not an option since it will very quickly become the norm and then compulsory by default. Third parties will increasingly want to use it as an identifier and so people will for forced to carry the "voluntary" card. For a truly voluntary card to be introduced there would need to be legislation to restrict its usage and even if this were the case it would be very difficult if not impossible to impose this restriction in practice.

> A universal entitlement card scheme would be one where: everyone in the country over a certain age was required to register with the scheme and to obtain a card; a card would be the only way to access particular services (other than in an emergency or in cases where a card had been lost or stolen).

This "universal" scheme is a compulsory scheme by any other name. The fact that every citizen must obtain a card makes it compulsory. The fact that the card is the only way to access services is further compulsion. There is a further concern that not carrying a universal entitlement card would arouse suspicion and suggest such people had some ulterior motive for doing so. Universality and social construction will ensure that carrying a card will become effectively compulsory.

It seems that either a voluntary or universal scheme will effectively breach a long-standing tradition in the UK that law-abiding citizens are not required to carry proof of their identity. It is therefore appropriate for Parliament to be given details of the full range of circumstances in which individuals will be required to produce such a card or report changes of personal details. This would help to maintain our civic tradition and also ensure the avoidance of function creep of the card.

According to the consultation document "a significant amount of personal information would be held in one place and there would need to be sufficient safeguards to prevent abuse." Given the talents and resources of criminals, and the potential rewards to them of obtaining access to such information, the security demands on such a system will be massive. The problem is that there is no such thing as a totally secure database. Indeed, the Government have been criticised for already using inherently insecure software and networks to deliver information.

A particular concern is that sensitive information is involved in many entitlements, such as entitlements to concessions (for example for disabled people) at leisure centres and swimming pools. While it is relatively easy to use an entitlement card to prove entitlement in such circumstances, the problem is how to still maintain appropriate confidentiality about this data, which is sensitive within the meaning of the Data Protection Act 1998. If there are sufficient controls over the release of sensitive data, the circumstances in which it is released to such facilities need to be strictly controlled, but it is impractical to maintain such strict controls over all of the thousands of leisure centres, which are run by diverse organisations.

Each additional type of use for a card is likely to raise similar issues (for example, entitlements to concessionary fares on public transport) and to increase the number of locations with card readers or database access terminals and the number of individuals with access to them that would have to be secured. As more people have access to this linked database the greater the likelihood of privacy violations. The only conclusion that can be drawn is that extensive use of a card to prove such entitlements is inevitably in conflict with practical maintenance of security of such sensitive data.

The card schemes as outlined in the consultation document will have severe costs and will not bring benefits as great as those claimed. It is inappropriate to introduce either a "voluntary" or a "universal" card scheme, since both are highly likely to become *de facto* compulsory card schemes of the sort that the Government has stated it does not wish to introduce.

3.16 Trusting the System [2003]*

This section was first published as: Rogerson, S. (2003) ETHIcol—Trusting the system. IMIS Journal, Vol 13 No 4, pp. 27–28. Copyright © Simon Rogerson.

At a recent Parliamentary Information Technology Committee (PITCOM) meeting the threat from cyber terrorism was discussed. Concerning our vulnerability, one speaker, Paul King from Cisco Systems explained that,

> I would estimate conservatively that if you are on the internet you are scanned every 20 seconds. The question is whether you notice. If they are scanning you, what is it for? The record I have for being scanned is an unclassified UK defence network which within four seconds of going live on the internet it was being scanned.

The second speaker, Dame Pauline Neville-Jones, Chairman of Qinetiq said,

> As we place more reliance on mobile computing, on electronic service delivery, as we move towards the so called pervasive computing paradigm, and as the availability of these technologies increases to a still wider proportion of the population, we shall find it considerably harder to protect against their misuse. All freedom in society is ultimately founded on trust—trust that in allowing you freedom to act, you will not harm my interests or curtail my freedom. Trust in the computing world is no different—indeed it is vital: the user must trust the system to be willing to use it on a repeated basis. New technologies may well increase the opportunities for the destruction of that trust.

Dame Pauline continued,

> the complexity involved in providing adequate security in the sort of computing environment I have described is of a different order from that faced today. Hence my caution in stating that while the threat from cyber terrorism is comparatively low today, the situation is not stable and is likely to be moving adversely. . . . The crimes will not have changed much in nature but the range of possibilities for committing them and the difficulty of protecting against them and tracking down the felons will increase manifold.

It seems we must all be vigilant against such threats particularly when they are against the very foundations of society. When asked about the potential risks from cyber terrorism to the electronic voting for the May local government elections and whether they should take place, the two speakers had different perspectives.

King replied, "We should definitely carry on. These are pilots and will be watched very closely, and it's only from projects like this that we will learn. I think they are safe because they are well resourced." Whilst Neville-Jones responded,

> Some basic things need dealing with and I don't know if that has been done. Apart from that I think the issue will surround how far and how fast you can go down this road. I would want to be more confident than I am now that the systems are not open to misuse before using them for a large scale election.

There is a groundswell of expert opinion that electronic voting using the internet is dangerous. Lee Dembart in the International Herald Tribune reported on 28 Aril 2003 that,

> In all electronic elections in Europe and most of the United States so far, security experts say, the systems used were vulnerable to attack and could have been manipulated in undetectable ways that would have made it impossible to determine that the results of an election had been changed, either by accident or design. Specifically, the experts say, internet voting could be crippled by a "denial of service" attack against the computer servers recording the vote, for which there is no known defense, and could disenfranchise large numbers of voters. In addition, they say, since voters use their own computers, election officials have no control over what software is installed on those machines or what viruses might be lurking in it that are activated only during an election to change votes.

Whilst Simon Parker writing in the Guardian on 30 April 2003 pointed out that,

> A security assessment carried out for the government suggests that, by making the voting system accessible across the globe, e-voting will "dramatically" increase the potential for trouble. Individual hackers, criminals, political activists and foreign intelligence services are among those who might try to rig the vote or destroy the technology used to run the election.

Finally, in their press release of 1 May 2003 the Foundation for Information Policy Research (FIPR) warned that electronic voting systems such as those being trialled in local government elections may lead to major problems and could severely damage the public's confidence in the electoral process.

FIPR warned that election integrity can be assured only if e-voting machines produce a paper audit trail that can be verified by voters and later by election scrutineers. This is not the case when votes are cast from telephones or insecure home PCs, as in the trials taking place in 18 local authorities during today's local elections. Voting machines must be squarely under the control of local election officials, who have a better chance of ensuring they are free from viruses or other malicious software that might monitor and corrupt a user's vote.

Without such precautions, FIPR claimed it will be impossible to prove afterwards that an election was carried out correctly. If problems occur, levels of public mistrust could make Florida voters worries about "hanging chad" look trivial.

There is a second important issue that of scaling related to the cyber terrorist and electronic voting. Lee Dembart explained that,

> Several experts noted that if people intended to rig an electronic election, they would not waste their time and effort on a minor local election with little consequence, thereby tipping off the authorities to the vulnerability of their election system. Such people would ignore small, pilot project elections, such as those currently under way, in order to increase the authorities" confidence in the system. They would wait until a big election, such as a national one, before attacking.

Simon Parker concurred with this view writing that,

> At De Montfort University's Centre for Computing and Social Responsibility, Leicester, researcher Ben Fairweather says: "I don't think we know for certain that an electronic

general election is possible at the moment. It might be possible, but one of the big problems is that piloting it at the local level you're not facing the challenges you'll face in the real thing." Few people, he suggests, would bother trying to rig or hack a local e-voting pilot, but a general election would be a far more tempting target.

Are these concerns well founded? In July 2003, the Electoral Commission will be publishing a review of all the pilots. Richard Allan MP has written a submission for the report concerning the internet pilot held in Sheffield. His submission seems to bear out the opinion of experts.

He wrote,

One polling station (Hunters Bar School, Broomhill Ward) never received its ISDN line installation. This meant that it could do no online checking all day and had to work from a hastily supplied paper register. There was no way to verify if people voting at that station had voted online or elsewhere meaning that anyone in that ward could vote twice—once at Hunters Bar School and once by any other method.

There were problems with several of the ISDN lines during the day. I encountered engineers at a couple of sites who told me they had been drafted in urgently as the main election contractors, BT, did not have the personnel available to deal with the call-outs. Each time an ISDN line went down no online checking could be carried out.

The most worrying result of the cumulative problems was the lack of security evident in this election in terms of ensuring that people only voted once.

All Broomhill residents could easily have voted twice all day. Most residents in other wards could have voted twice by visiting a polling station that was not performing checks at some point in the day.

People who had no right at all to vote in a ward election could have done so by going to a polling station with no online checking and giving any reasonable sounding name and address.

I am confident that these failures would have been sufficient to lead to a challenge to the result if any party had lost by just a few votes in this election.

The message is clear when using technology to support the fabric of society, we must be cautious and ensure that cyber terrorism threats are effectively counteract to the satisfaction of the citizens of society.

3.17 What is Wrong with Mobile Phones? [2003]*

This section was first published as: Rogerson, S. (2003) ETHIcol—What is wrong with mobile phones? IMIS Journal, Vol 13 No 5, pp. 25–26. Copyright © Simon Rogerson.

Cell phones have become one of the icons of modern living symbolising a world of the instantaneous, of the connected and of the disposable. But behind the iconic triviality lie serious issues which affect individuals and society alike.

Cell phones have become the ultimate designer fashion accessory with costly price tags. There is both marketing and peer pressure, particularly on the young, continually to update their phones in order to keep up with trends. This is socially divisive. There is also an associated environmental issue. The average shelf life for a cell phone is currently 18 months. By 2005 it is estimated that 130 million cell phones will be thrown away annually representing 65,000 tons of waste a year. This is an environmental hazard.

On the one hand these new instruments of communication demand increased levels of literacy and technical literacy skills. On the other hand, they are having significant impact on the use of language. We are seeing a simplification of language which endangers our linguistic culture and heritage, and results in a loss of nuance, meaning and subtle shades of difference.

A number of health issues need to be addressed. There is a contradictory literature concerning microwave transmissions from handsets and ground stations. This is particularly concerning regarding children. Small keypads can cause problems for those with limited dexterity. There is some evidence to suggest repetitive strain injury is a problem for those who frequently send text messages. Finally, the use of cell phones and text messaging in particular can become a compulsion or even an addiction.

Trends in use raise some interesting issues. Carrying active cell phones provides a mechanism for surveillance and tracking by third parties. As we increase the use of our cell phones we become more vulnerable to receive a new form of spam—the junk text message. This is becoming an increasing problem.

Using cell phones (even with hands-free facilities) whilst driving presents new dangers. A driver's concentration is diverted to the conversation with the person on the phone. This is different from conversation with in-car passengers as in this situation both driver and passenger are aware of road conditions and temper their conversation accordingly. Given the "street value" of cell phones, users are increasingly at risk from mugging when using phones in public spaces. The use of mobiles in public spaces raises another issue. Such conversations intrude into other "quiet spaces" and infringe on the privacy of others. This has led to a new concept of "mobile free zones" on trains. There is increasing pressure for us to remain in mobile contact when away from the office. The electronically-enabled culture of instantaneous response to the demands of employers and clients has become the norm. We can no longer leave work at the office.

But impacts are not always obvious and direct as illustrated by this extract from the autumn 2002 online edition of *Seeing is Believing*.

Cell phones may have revolutionized the way we communicate, but in Central Africa their biggest legacy is war. Nearly 3 million people have died in Congo in a four-year war over coltan, a heat-resistant mineral ore widely used in cell phones, laptops and playstations. Eighty percent of the world's coltan reserves are in the Democratic Republic of Congo. The mountainous jungle area where the coltan is mined is the battleground of what has been grimly dubbed "Africa's first World War," pitting Congolese forces against those of six neighbouring countries and numerous armed factions. The victims are mostly civilians. Starvation and disease have killed hundreds of thousands and the fighting has displaced 2 million people from their homes. Often dismissed as an ethnic war, the conflict is really over natural resources sought by foreign corporations— diamonds, tin, copper, gold, but mostly coltan. At stake for the multitude of heavily armed militias and governments is a cut of the high-tech boom of the 1990s, which sent the price of coltan skyrocketing to peak at US$400 per kilo. Coltan—short for Colombo-tantalite—is refined into tantalum, a "magic powder" essential to many electronic devices. The war started in 1998 when Congolese rebel forces, backed by Rwanda and Uganda, seized eastern Congo and moved into strategic mining areas, attacking villages along the way. The Rwandan Army was soon making an estimated US$20 million a month from coltan mining. A May 2002 report from the United Nations Security Council said the huge coltan profits are fuelling the war and allowing "a large number" of government officials, rebels and foreigners "to amass as much wealth as possible." The fighting rages on despite peace treaties signed in the summer of 2002.

Such technological developments as cell phones need to be assessed for potential risks and benefits. The identification of risks then requires effective action which might include development modification and instruction in proper use of those for which it is intended. Overall, we must always strive to take a balanced view of technological advances and potential of these amazing human endeavours.

3.18 Market~ing Forces [2004]*

*This section was first published as: Rogerson, S. (2004) ETHIcol—Market~ing Forces. IMIS Journal, Vol 14 No 1, pp. 31–32. Copyright © Simon Rogerson.

The world of marketing has always struggled to establish a favourable reputation in the eyes of the public. Whether it is reasonable for the public to be wary of all marketing is debatable. What is certain though is that it only takes one thing to damage a favourable or improving reputation to the point where that reputation may never be recovered. This edition of ETHIcol discusses two examples of questionable IT-enabled practice in the world of marketing which appear to hold marketing in an unfavourable light.

The first example concerns the marketing organisation Metronomy. Information on its website (www.metronomy.com) explains that, "Metronomy is an innovative desktop marketing concept that aims to bring TV quality advertisements into peoples' homes via a PC. This allows advertisers to show fully interactive TV-quality adverts to specifically targeted households." The company states that it will give "households a free IBM PC for 3 years in exchange for their pledge to watch up to 3 minutes of advertising per hour on the PC screen." This is a commitment to 1,080 hours of viewing.

Details about the household and its members must be submitted to Metronomy as part of the application process. Households must agree to connect to the internet at least once per month and must subscribe to an Internet Service Provider (ISP) for 36 months. A monthly update of advertisements will be sent to households each month and must be installed within seven days in order for the PC to remain operational.

PCs will be monitored to ensure households fulfil their advertisement viewing obligations. Other details will not be monitored. All users' personal information will be kept confidential. A monthly update confirming household viewing patterns will be automatically sent to Metronomy. This will be used together with consumer type and location data to tailor the advertisement updates on the CD ROM sent each month to households.

There are many issues raised by this new form of interactive marketing directly into the home. Vulnerable people are likely to be attracted by the lure of a free PC without realising the hidden costs of committing to 36 months of ISP subscriptions and the need to insure the PC. It is unclear what happens to the information either supplied by households or collected automatically. It is unclear what data is recorded about PC usage. There is no clear information that advertisements will be targeted based on collected information. There is no stated privacy policy. There are no published policies for handling complaints or dispute resolution. There are no details of how breaches in contract by the householder will be dealt with.

In an article written by Jo Best for silicon.com, John Thornhill, chief executive of Metronomy is reported as saying,

> Metronomy believes that the offer of a free PC will benefit millions of UK households. We are delighted to be working with some of the world's leading media and technology companies including IBM, Omnicom and Interpublic, whose support has been instrumental in developing this ground-breaking initiative.

The implication of this being a social good is very misleading. This is simply a new form of marketing attractive to many suppliers of goods and services who can see the benefit of using enticing and interactive advertising directly into the homes of specifically targeted households. Metronomy have a responsibility to ensure people fully understand the proposition that is being offered. Currently the information supplied on their website falls way short of fulfilling this obligation. The one thing households should remember is there is no such thing as a free PC!

The second example concerns the use of digital images to market products and goods. In a recent article for the Guardian, Sean Dodson discusses the latest generation of digital images of females and their sense of realism with the inclusion of human blemishes. So realistic are these images that they are now being used as a marketing tool.

Dodson writes,

> On television adverts, in movies and, very soon, on your mobile phone, the use of ever more complicated digital models is becoming more commonplace. In Germany, the design studio NoDNA is populating European interactive TV channels with a procession of virtual presenters. In France, the digital model Eve Solal has been signed by the Ford model agency and she even has her own Saturday morning radio show. Closer to home, the DA group of Glasgow produce what it describes as a range of "interactive agents." It, too, has a virtual pop star in the form of Tmmy (pronounced "Timmy") and it has also recently created Seonaid, an online news presenter for the Scottish Executive. The company's next plan is to bring digital models to mobile phones with a range of avatars that will perform instant messaging tasks. In Japan, digital models have been used to sell anything from cosmetics to computers to cash loans.

Whilst animation of digital models is still in its infancy, static images are often mistaken for real people; for example, a group of 100 students were unable to distinguish correctly between real and digital images in set of 12 pictures. What is certain is that digital models will be increasingly used in the marketing of all sorts of products and services. The concern is that these images are always of females suitably created to have maximum appeal. They are designed to prey on our desires and egos. It is only a matter of time before the animated digital model will be indistinguishable from the real form. How will they be used by marketing to entice us to purchase goods and services that we perhaps neither need nor can afford?

The two examples illustrate how marketing can use IT to increase its power and influence over the consumer. The balance between the acceptable and the unacceptable has become more acute. The responsibilities and obligations of marketing in the information age need to be redefined and accepted by all those working in the area.

3.19 Tag Ethics [2004]*

*This section was first published as: Rogerson, S. (2004) ETHIcol—Tag Ethics. IMIS Journal, Vol 14 No 5, pp. 31–32. Copyright © Simon Rogerson.

Radio frequency identification (RFID) is heralded by many as one of the new society-changing technologies. The RFID tags are minute ranging from as large as a grain of sand to as small as a speck of dust. This intelligent Lilliputian technology has a huge range of applications. Tags can be placed in absolutely everything. The current list includes animal identification, beer keg tracking, vehicle key-and-lock anti-theft systems, library book or bookstore tracking, pallet tracking, building access control, airline baggage tracking, clothing item tracking, and identification badges.

Electronic toll collection such as the FasTrak system in California use RFID tags. As the vehicle passes the tag is read and the information is used to automatically debit the toll from a vehicle owner's bank account. Retailers such as Wal-Mart and Tesco are using tags in their supply chains to track and monitor items from wholesale supply to public consumption. Manufacturers such as General Motors use the tags to monitor and report where every item is at every moment during the manufacturing process. Recently the Ohio Department of Rehabilitation and Correction in the US has announced it will fit RFID transmitters the size of a wristwatch to all inmates so they can track them all the time. If prisoners try to remove the tags an alert will be sent to the prison computer system and the alarm will be raised. Finally, the US military has recently announced it will tag all its assets as part of a major overhaul of logistics and security.

The potential is huge but the price we have to pay may be high. Many argue that the depth of information which can be held by these tags, the ease with which they can be incorporated into products and the ability to interrogate them at a distance present major issues for society particularly regarding privacy.

TheFreeDictionary.com (encyclopedia.thefreedictionary.com/rfid) lists four main privacy concerns regarding RFID technology as:

- The purchaser of an item will not necessarily be aware of the presence of an RFID tag or be able to remove it.
- An RFID tag can be read at a distance without the knowledge of the individual.
- If a tagged item is paid for by credit card or in conjunction with use of a loyalty card, then it would be possible to tie the unique ID of that item to the identity of the purchaser.
- RFID tags create, or are proposed to create, globally unique serial numbers for all products, even though this creates privacy problems and is completely unnecessary for most applications.

These align with the reported problems over RFID tags and readers described by Technovelgy (www.technovelgy.com/ct/Technology-Article.asp?ArtNum=20#Security).

- The contents of an RFID tag can be read after the item leaves the supply chain.
- RFID tags are difficult to remove.
- RFID tags can be read without your knowledge.
- RFID tags can be read at greater distances with a high-gain antenna.
- RFID tags with unique serial numbers could be linked to an individual credit card number.

Larry Ponemon (2004) warns of another potential problem with RFID tags. He explains that over-reliance can lead to complacency in stringent monitoring in the supply chain which can have devastating effects in for example food manufacturing. He explains that there is a "need to recognise the importance of having internal controls such as monitoring and accountability procedures in place to identify negligence in the supply chain or improper usage of personal data."

These commentaries illustrate the range of issues that need to be addressed if RFID technology is to be widely accepted.

Some organisations have been mindful of public concern over the seemingly unrestricted used of RFID tags. For example, in its 2003/2004 Corporate Social Responsibility Report Marks and Spencer states,

> We started to introduce [RFID] technology into our food distribution systems in 2002 to increase the efficiency and accuracy of delivering products to stores. . . . However, a number of civil society groups have concerns that it could then be used unethically to track people and their behaviour. We held discussions with the National Consumers Council and Caspian to understand their worries and have developed responses for a number of their concerns. For example, any RFID tags used on products are made very visible and easily detachable. It is our intention to continue these discussions. . . . We see great potential in RFID technology and are committed to ensuring we use it in a way that is acceptable to our customers and wider society.

Speaking at the RFID Privacy workshop held at MIT in US, Wipro Technologies, a leading IT solutions and services provider in application development, system integration, product implementation, and consulting services, called for improved public policy concerning RFID technology. This would help the commercial use of RFID realise its full potential, address the radical shift to a customer-oriented marketplace, address the issues surrounding reengaging with customers, address the growing complexity of conducting business globally and improve the trust in public and private institutions. An integrated approach to public policy was called for based on six elements: technical, industry self-regulation, ethical approach, legislation, RFID branding, and consumer education. The ethical element would be based upon:

- respect confidentiality
- don't "flame"
- don't be anonymous
- don't allow third party to access other's data
- don't misrepresent or lie
- follow government's general guidelines
- consider presentation of message

This strong ethical element is encouraging particularly as it is being put forward by a major player in the computer industry. It remains to be seen whether such an integrated approach to international public policy will ever come to fruition.

This pervasive technology has great appeal to many organisations that have a legitimate requirement to track and monitor goods and people. But for it to go unfettered is dangerous. How do you know that the item of clothing you are wearing has not got a tiny RFID tag in it which has not been deactivated and now someone knows where and when you are reading this article?

3.20 A Lesson from History [2006]*

*This section was first published as: Rogerson, S. (2006) ETHIcol—A Lesson from History. IMIS Journal, *Vol 16 No 2, pp. 29–30. Copyright © Simon Rogerson.*

Recently I had the privilege to accompany a group of school children on a trip to Auschwitz and Birkenau. By the end of the war some six million Jews and many millions of Poles, gypsies, prisoners of war, homosexuals, mentally and physically handicapped individuals, and Jehovah's Witnesses had been murdered. The visit was a time of humble reflection and the placing of things in context. The historical account of human suffering is sickeningly shocking but alongside this is the realisation of the evil brilliance not mindless thuggery that orchestrated the "Final Solution."

The Scientific Management Principles (1911) of F W Taylor promote efficiency in an industrial process. Such principles appear central to the attempted extermination of a race using the abhorrent industrial processes at Auschwitz and Birkenau. For example, testimony of engineer Fritz Sander (7 March 1946) states,

> This [crematorium] was to be built on the conveyor belt principle. That is to say, the corpses must be brought to the incineration furnaces without interruption. When the corpses are pushed into the furnaces, they fall onto a grate and then slide into the furnace and are incinerated. The corpses serve at the same time as fuel for heating of the furnaces.

So what does this have to do with computing? What if these events were taking place today? Technological Determinism argues that technology is the force which shapes society. Computing power would therefore be a major force in activating the "Final Solution." Michael Porter's Value Chain Analysis (1985) is one way to consider the impact of this force. Here are just a few examples. These examples are based on computer application systems that exist today and which are proven and accepted.

Inbound Logistics: the receiving and warehousing of raw materials, and their distribution to the industrial process as they are required.

Computerised transportation scheduling can minimise cost and ensure timely delivery to the points of industrial process. Humans are the raw material of this particular industrial process. Scheduling would enable enormous numbers of humans to be moved across occupied Europe in an efficient and timely manner. The effective flow of raw materials is a key factor in computerised industrial processes such as just-in-time manufacturing. The arrival of humans could be controlled by calculating transportation routes and speed so that there was a steady flow which did not overwhelm the camps or the industrial process. Rerouting and readjustment of speed could be triggered by "production data" being electronically communicated from the camps and industrial process.

Operations: the processes of transforming inputs into finished products and services.

Computerised process control is a method for maximising throughput, minimising disruption and facilitating non-stop processing. The input flows of both human and chemical raw materials of the gas chambers could be fully automated to increase throughput. Once dead, humans need to be moved. The use of robotic devices would enable mounds of corpses to be loaded on to computerised conveyor belts which would route bodies to the next available furnace. This process would be endless.

Automatically controlling the flow of corpses would open the possibility of secondary processes that are alternatives to cremation. For example, the element phosphorus is relatively rare in nature, yet it is of vital importance to life. Bone meal is often used as a supplement for calcium and phosphorus through, for example, fertilizers. What would stop the "Final Solution" having two outcomes cremation and fertilizer manufacture?

Outbound Logistics: the warehousing and distribution of finished goods.

The sorting of personal possessions, reuse of personal possessions, and the recycling of materials could be facilitated by computerised warehouse control systems and goods delivery systems.

The Infrastructure: organisational structure and control systems.

The annihilation of sectors of the population can only succeed if it is founded on meticulous record keeping which identifies and tracks every member of a given sector. This is a manual impossibility but with computers is alarmingly easy. The linking of biometric identity tagging with genetic/DNA birth records provides the means to identify anyone. An individual's identification and location in computerised form enables inbound logistics and operations to identify and use all desired humans. Some could be redirected to slave labour camps before becoming the raw material of the "Final Solution" industrial process.

This account might be shocking to many readers. That is its intention. It seems that if the Holocaust had occurred in our technologically advanced modern world there is a very good chance that it would have completely succeeded. If ever there was an example to convince us, as custodians of the most powerful technology yet devised, of our responsibilities and obligations to humankind, this is it. To the narrow-minded technologist, who seems intent on viewing the world as a computer playground where anything is possible and everything is acceptable, it is time to act with professional responsibility. Remember, "All that is necessary for the triumph of evil is that good men do nothing" (sometimes attributed to Edmund Burke).

Note: The Nazis were aware of the power of technology. Joseph Goebbels, the Reich Minister of Propaganda of Nazi Germany, gave a speech about the power of radio in political dominance in which he explained that,

> It would not have been possible for us to take power or to use it in the ways we have without the radio and the airplane. It is no exaggeration to say that the German revolution, at least in the form it took, would have been impossible without the airplane and the radio.
> (translation source: http://research.calvin.edu/ german-propaganda-archive/goeb56.htm)

3.21 Let the Games Begin [2006]*

*This section was first published as: Rogerson, S., Gittings, C. & Lapper, J. (2006) ETHIcol—Let the Games Begin. IMIS Journal, Vol 16 No 3, pp. 31–32. Copyright © Simon Rogerson, Chris Gittings & Johnathan Lapper.

It is generally recognised that one of the forefronts of computing lies within games technology. The commercial incentives to produce "top 10" games are enormous. But just like the music industry fame is short lived and once more the quest is on to produce yet another top offering. It is this that drives the technology forward in an attempt to recreate realism in every way and so produce even more evocative games. These technological advances are heralded by industry and adapted to new applications outside the entertainment industry. This is a rich technological vein to be mined but are there any detrimental costs?

The range of games is vast. They can be split into genres such as Action including Platform, Fighting and Hack and Slash, Sport, Strategy, Role Playing, Adult and Mature, Driving, Simulation,

Activity, Classic, and Educational. A game may have more than one genre. There seem to be competing requirements associated with any game's development project. Typically, a game should:

- cater for a wide variety of tastes
- run on multiple platforms
- sell at competitive price
- be affordable
- incorporate reward scheme for players
- motivate to continue playing
- be entertaining
- be socially acceptable
- not damage health

The implications of these requirements on various stakeholders could be profound. For example, children could be encouraged to spend all their pocket money (and more) on games to satisfy their widening interests of or even addiction to a particular game genre. Some games may be considered by society to be unacceptable because of the inclusion of violence, sexually explicit acts, criminal acts or other antisocial behaviour. Acceptability has a cultural/regional dimension which affects the way in which vendors must operate. For instance, one easily accessible adult game states that it is "the ultimate adult sex game. It contains thousands of hot adult interactions and features intelligent game play which matches interactions to the player status and the state of the game." It does however have a locking function which "ensures that minors do not intentionally or accidentally view any of the content of this game." Whether the inclusion of such a lock makes it more societally acceptable is open to debate.

A recent interview with a game developer revealed an interesting perception on the professional role of games technologists. The interviewed developer described at length the roles of five necessary categories of developers: Lead Programmer, Games Programmer, AI Programmer, Tools Programmer and Engine Programmer. All had a challenging set of technological responsibilities but none had any which related to the impact on stakeholders such as society, parents, children, and teenagers. Indeed, when asked if there was an awareness of the broader social issues associated with games technology the response was to focus on the gaming fraternity and the developer commented that

> One of the most important developments within the last decade of gaming is online gaming. Before this gaming was considered as a pretty anti-social activity, something that you would do by yourself at home. But with the internet and home consoles online play gaming has become a much more social activity.

It seems the technologists are oblivious of or do not want to face up to some difficult ethical and social issues surrounding games technology.

As games become more realistic do they warrant even more careful scrutiny before release? For example, the very popular Hack and Slash genre of game requires that the player goes around the levels hacking and slashing through the enemies using different weapons and techniques. As the player proceeds through the levels the weapons or abilities become stronger but the enemy's weapons also become stronger. It is claimed by vendors that such games appeal to those who want a lot of fast action as there will always be something to fight around the next corner. With improving graphics and animation, will the realism of the game environment have a different psychological effect on the game player than earlier generations of games? Does the game player still separate the virtual world of the game from the reality of his/her own life?

One of the recent advances in games is the development of hybrid games which combine physical action and virtual action. There are currently two broad social categories on offer. There are those which can be described as innocent fun. For example, there are quiz games with interactive buttons, dances games which require physical copying of computer-generated dance steps, and singing related games. But there is a second more sinister category which centres around physical combat. The game player no longer uses the console to engage in combat but uses a physical mock-up of a weapon such as a gun or chainsaw or performs martial art kicks against a pole. These physical interactions are then linked to the game environment and the game player sees the immediate effect of the physical effort translated into the virtual world. There appears to have been no thought given to the blurring of the boundary in these combat hybrid games. This is concerning as we seem to be entering into an unknown world which might detrimentally affect the game player. Indeed, might this have a knock-on effect on society?

There will be those who question this questioning. The benefit of being able to blur the physical and virtual has so many potential applications in business, education, health and communication. If hybrid combat games fund this advance then the ends justifies the means—or does it?

3.22 Ethics of the Blogosphere [2006]*

This section was first published as: Rogerson, S. (2006) ETHIcol—Ethics of the blogosphere. IMIS Journal, *Vol 16 No 5, pp. 32–33. Copyright © Simon Rogerson.*

The converging technologies have provided society with new ways in which to communicate. Today social software is opening up new opportunities for interaction as well as an antidote to establishment and mainstream communication. Weblogs or blogs are at the forefront of this cultural revolution.

According to Wikipedia,

> A blog is a type of website where entries are made (such as in a journal or diary), displayed in a reverse chronological order. [Each posting is usually time-stamped.] Blogs often provide commentary or news and information on a particular subject, such as food, politics, or local news; some function as more personal online diaries. A typical blog combines text, images, and links to other blogs, web pages, and other media related to its topic. Most blogs are primarily textual although some focus on photographs (photoblog), videos (vlog), or audio (podcasting), and are part of a wider network of social media.

A new blog is created every second!

Martin Kuhn suggests there are two types of blogs—personal and journalistic. A personal blog is typically maintained by an individual as a personal electronic diary. Whilst a journalistic blog focuses on information provision, media watchdog roles and breaking original stories. Some blogs which attract large readerships become commercial selling advertising space.

Whilst bloggers argue that freedom of expression is the fundamental principle of the blogosphere nevertheless each blogger does have responsibilities to society in general simply because the blog is quickly becoming one of the most used strands in society's communication web. But there is a problem. As Rebecca Blood explains,

> the very things that make [blogs] so valuable as alternative news sources—the lack of gatekeepers and the freedom from all consequences—may compromise their integrity and thus their value. . . . There has been almost no talk about ethics in the weblog universe: Mavericks are notoriously resistant to being told what to do.

In 2002 Blood suggested six principles of ethical behaviour for bloggers.

1. Publish as fact only that which you believe to be true. If your statement is speculation, say so.
2. If material exists online, link to it when you reference it. Linking to referenced material allows readers to judge for themselves the accuracy and insightfulness of your statements.
3. Publicly correct any misinformation.
4. Write each entry as if it could not be changed; add to, but do not rewrite or delete, any entry.
5. Disclose any conflict of interest.
6. Note questionable and biased sources.

These ideas have been discussed on several blogs and have formed the foundation of further work undertaken by Martin Kunn in 2005. Kuhn put forward a Code of Blogging Ethics which provides a framework for both personal and journalistic blogs whether they are amateur or commercial. Here is the proposed code.

- Promote Free Expression.
- Be as transparent as possible as to personal biases and affiliations.
- Emphasise the "human" element in blogging.
  - Reveal identity.
  - Promote equality in the blogosphere.
  - Minimise harm to other.
  - Actively promote community building.
- Prioritise factual truth.
  - Never intentionally deceive readers.
  - Be accountable for information posted on your blog.
  - Cite and link to all sources.
  - Secure permission before linking other blogs or web content.
- Promote Interactivity.
  - Post regularly to your blog.
  - Respect blog etiquette and protocol.
  - Be entertaining and interesting.

There is much to be gained from these two sets of guidelines. Indeed, communication in general would be better if such ideas were adopted by us all. In his work Kuhn posed several questions to bloggers.

- Who are the stakeholders with regard to your blog?
- Who will be affected by what you post?

- When making decisions about your blog, do any of the following values or duties cross your mind: transparency, accountability, minimising harm to others, free expression, factual truth, and etiquette? If so, which? Can you rank them?
- Are there any other values/duties you feel should be weighed in a discussion of blogging?
- Are there certain duties all bloggers should fulfil, all of the time, in order to be good bloggers?
- Are there certain things bloggers should never do?
- On a societal level, what role do blogs play?

Most readers of this account will have accessed blogs at some point. Try to answer these questions. Why not post your ideas on the Web so that the online community becomes more aware of these issues and starts to takes ownership of promoting acceptable conduct in online citizens.

3.23 Voicing Concern [2008]*

*This section was first published as: Rogerson, S. (2008) ETHIcol—Voicing Concern. IMIS Journal, Vol 18 No 2, pp. 24–25. Copyright © Simon Rogerson.

Voice over Internet Protocol (VoIP) is a protocol that allows basic communications functions, such as voice calls, voice messaging and facsimile over the internet instead of the traditional telephone network. It is hard to find data on VoIP usage but it is suggested from several sources that around two billion people have used this internet facility. This article focuses on one of the many VoIP service providers, Skype. It is based on experiences as a first time Skype user with a keen interest in service integrity!

Loading the free software for Skype is very easy and within no time you can be a registered Skye user communicating with people around the world. However, unbeknown to the new user are the default values on installation. On booting up your computer, Skype opens automatically. Why might this be a problem? It is because of the default privacy settings on installation which are "allow calls from anyone," "automatically receive video from anyone," "allow chats from anyone," "allow my status to be shown on the web," "accept Skype browser cookies" and "keep chat history forever." This maximum visibility makes Skype users vulnerable. Is it acceptable for a VoIP service provider to operate an "opt-out" strategy rather than an "opt-in" strategy and give no apparent warning of this? Once a connection is made and a web camera is activated the recipient can see you and can capture still images and video streams without your knowledge or consent. Once captured these can be easily manipulated and distributed. This potentially covert capture of visual information increases Skype users" vulnerability as such information can be used in a number of unsavoury and questionable ways.

It is likely that a Skype user will want to portray himself or herself as someone whom people would like to talk to. Skype allows you add facts about yourself such as full name, date-of- birth, gender, home location and narrative about yourself, for example, where you could describe your social and sporting interests and your favourite music. All these details will always be seen by all Skype users. How many Skype users will consider that adding too much personal data increases your vulnerability? Skype also allows you to share personal data with the social networking site MySpace. Such interconnectivity further increases vulnerability.

Skype is supported by a website which enables Skype users to manage and pay for services. Skype's privacy policy, which was last revised on 8 September 2007, is published on this website.

(www.skype.com/intl/en-gb/legal/privacy/general/ accessed 17 March 2007) The policy raises a number of ethical issues. Consider these extracts, each of which is followed by an issue that they raise.

"Skype collects and processes, or has third party service providers acting on Skype's behalf collecting and processing, personal data relating to you"—It is unclear who are the third parties and how such organisations are vetted regarding privacy of Skype users.

"Your information may be stored and processed in any country in which Skype and the Skype group maintain facilities, including outside of the EU."—It is not mentioned that there are some countries which the EU will not allow personal data to be transferred to due to that country's unacceptable track record on personal data. What would Skype do if it operated in such a country?

"personal information of Skype users will generally be one of the transferred business assets. We reserve the right to include your personal information, collected as an asset, in any such transfer to a third party."—Skype clearly views personal data as a business asset which can be sold as part of company sale. What safeguards do Skype users have regarding the use or abuse of their personal data in such situations?

"Cookies enable Skype to gain information about the use of its websites. This information may be analyzed by third parties on our behalf."—Once again it is unclear who the third parties are and how they have been vetted.

"Skype websites may contain links that will let you leave Skype's website and access another website. Linked websites are not under the control of Skype and it is possible that these websites have a different privacy policy."—As a marketing strategy Skype has established these links to enhance the attractiveness of Skype to potential users but seems to have abdicated any responsibility about the integrity of the websites it has exposed Skype users to.

"Skype and, where relevant, the Skype group entities will retain your information for as long as is necessary"—Given Skype views such personal information as a business asset would imply that data is kept forever.

The Skype entry on Wikipedia (http://en.wikipedia.org/wiki/Skype#Security_concerns accessed 16 March 2008) includes a number of associated concerns about privacy policy and security of Skype communications. It states,

> Skype was also found to access BIOS data to identify individual computers and provide DRM protection for plug-ins. It cannot be assured that Skype calls are not interceptable. Skype provides end to end encryption for connections between users however in an interview at cnet.com Skype chief security officer Kurt Sauer would not eliminate this possibility. Skype is owned by eBay, whose privacy policy is perhaps the most liberal of any large corporation—eBay claims it goes above and beyond what it is required to do by law, seeking out and giving police all the information it stores about users excluding some financial data, for which they require a subpoena

VoIP services are a relatively new example of ways in which we can use the internet to link up with people round the world. Users of such services have an obligation to ensure they are careful in how they use such services and do not put themselves at risk. Equally the service providers have an obligation to minimise user vulnerability through clear advice and through integrity of their services and associated policies. It would seem there is a long way to go before such obligations are understood and acted upon.

3.24 Ethics of the Street [2009]*

This section was first published as: Rogerson, S. (2009) ETHIcol—Ethics of the Street. IMIS Journal, Vol 19 No 3, pp. 27–28. Copyright © Simon Rogerson.

Google Street View is a facility within Google Maps and Google Earth that provides 360° horizontal and 290° vertical panoramic images of streets. It was launched in the United States on 25 May 2007 and now covers many towns and cities in United States, United Kingdom, Netherlands, France, Italy, Spain, Japan, Australia, and New Zealand. Google suggests with Street View,

> you can easily find the exact location for your crucial meeting and the nearest coffee shop, plan a restaurant venue to meet friends for dinner, or find the best viewing spot for a marathon or parade. If you are moving house or re-locating, you can save time by exploring properties and their surrounding area in advance and also by looking up driving directions. Street View allows you to check out a hotel or holiday home before you book, and to explore different travel destinations around the world.

There have been mixed reactions to Street View around the world. There are many supporters. For example, the BBC reported that Tate have worked with Google to integrate precise locations in the UK associated with artworks by JMW Turner and John Constable, which can then be viewed alongside their real-world locations. There have been numerous complaints about invasions of privacy. In Japan complaints that Street View cameras were mounted so high that pictures were taken over private fences and into homes has led to Google lowering cameras by 16 inches and retaking streets pictures. However, for example, the Information Commissioner's Office in the UK ruled in 2008 that the face-blurring and licence plate-blurring were sufficient to ensure that privacy was maintained.

The Google policy on personal privacy states,

> Street View only features photographs taken on public property and the imagery is no different from what a person can readily see or capture walking down the street. Imagery of this kind is available in a wide variety of formats for cities all around the world. We are committed to respecting local laws and norms in each country in which we launch Street View. Blurring technology and operational controls like image removal are amongst the ways in which we ensure that an individuals" privacy is respected. We make it easy for users to ask to have photographs of themselves, their children, their cars or their houses completely removed from the product, even where the images have already been blurred.

A cursory tranche of Street View in Coventry, UK revealed many vehicles with the registration plates not blurred and many people with their faces visible. It is true that Google provide a mechanism to rectify this. The visible plate JxxxPxx of a car in the drive of 35 Wrigsham Street was reported and information requested as to why this had not been blurred. Four weeks later, the plate is now blurred but there has been no communication to confirm this or a response to the question.

Is it reasonable to have a system which is not accurate in blurring images and which places the burden on the public to report flaws and request such flaws are rectified? Is it reasonable that there is no communication to complainants that such flaws have been rectified? It is uncertain whether the recording of pictures from a moving camera with surround vision high in the air is the same as a person standing in the street or even on the top of a double-decker bus. Google claims it is. However, the former is a comprehensive permanent view that can be revisited whilst that latter is a temporary view with restricted field of vision. Google's action to remove images following a complaint comes too late because many of these images have often been captured and presented on websites explicitly set up to show Street View images. Some images have become infamous. These include images of an old woman inadvertently exposing herself, men and women urinating in the street, naked sunbathers, a young male, who is possibly a burglar, climbing into a window of a house, and people entering and leaving adult entertainment establishments. These are data shadows which are impossible to remove as they have now seeped across the web. Whatever harm has been caused by such images cannot be reversed.

Whilst Google identify legitimate uses for Street View in "Top 10 Street View tips for everyday use" and "Top 10 Street View tips for businesses and organizations" there are also unacceptable potential uses that can be made of it. For example, it is easy to identify houses which have burglar alarms and windows open which can then be used to draw up a short lists of burglary targets. Areas where children frequent can be identified and even with face blurring some children can be recognised both of which increase risks to child safety. Currently towns and cities are being photographed but when small conurbations of just a few houses are photographed then lack of anonymity increases. Images of people in embarrassing situations could lead to harassment.

The technology behind Google Maps, Google Earth, and Google Street View is impressive. It provides an amazing interactive image of our world. However, it is an uncontrollable image because of the way in which flaws and complaints are handled retrospectively.

3.25 Ensuring Ethical Insurance [2009]*

This section was first published as: Rogerson, S. (2009) ETHIcol—Ensuring ethical insurance. IMIS Journal, Vol 19 No 5, pp. 28–29. Copyright © Simon Rogerson.

At the recent Computer Ethics Philosophical Enquiry, Oliver Siemoneit presented a paper *Ethical Issues of Pervasive Computing in the Insurance Industry*. This edition of ETHIcol discusses his presentation.

Advances in nanotechnology, microelectronics, and communication technology are characterised by the miniaturisation of ICT components which are cheap to produce and can be embedded in everyday objects. These embedded ICT systems are locally and globally interconnected using wireless technology. They collect data about their environment and adapt their behaviour according to that data as well as transmit tat data onwards. The claim for this pervasive computing is that clothes, cars, building, and even human bodies, for example, can be invisibly equipped with embedded ICT systems which collect, receive, process, save and communicate data, which in turn will ease our life and open up new opportunities for us.

It appears insurance companies have started to use pervasive computing in order to offer new products to their customers. This is based on two approaches. By using pervasive computing applications real-time risk-relevant data can be collected which enables insurance companies to calculate more accurate insurance premiums based upon the actual size of the risk and the probability of it occurring. Early warning and detection systems can be embedded which can prevent damage or reduce the amount of the loss.

For example, some motor insurance companies offer vehicle trackers which collect risk-relevant data such as driver details, trip duration, break times, driving characteristics, speed, vehicle location, and road conditions. This data is then used by the insurance company to build a profile of the vehicle use, estimating the likely risks which leads to raised or lowered premiums. In an attempt to make such pervasive computing applications more acceptable, they are often linked to additional services such as emergency location of vehicles and traffic congestion management.

It is not just inanimate objects that can be monitored. For example, sensors in sports shoes could be used by a partner company to devise and control training programmes. The collected data could be shared with the health insurance company where premiums could be adjusted to reflect the efforts of a client to stay healthy. The same sensors could be used to alert emergency service of dangerous values of life critical parameters.

It is the loss of privacy that seems to be one of the main ethical problems with this type of ICT application. In the process of collecting risk-relevant data there is the by-product of establishing a detailed profile of an individual's activities, preferences, and habits. Such profiles are very sensitive personal data sets. Those wishing to be insured may well underestimate the loss of privacy involved and indeed may be ignorant of this all together. The concept of informed consent in such circumstances appears to have been given insufficient consideration. Ironically the drive to manage risk by the insurance company creates a new risk to the insured. There is a clear tension between privacy and financial goals.

There is a second ethical issue that needs to be considered. The profiling of individuals by managed insurance companies has the potential to enable cost savings, but may enable discriminatory or exclusionary effects at the same time. This can run counter to the ethical principle of nonmaleficence for some, even while promoting beneficence for others. Indeed, some individuals, through no fault of their own, may find themselves uninsurable with the use of pervasive computing to collect risk-relevant data whereas previously the cost was spread widely across the whole insured population. There seems to be a difference between these individuals and those who have consciously caused themselves to be a greater risk to insure.

A suggested way of addressing such issues is the manner in which data is collected using pervasive computing. There could be a reduction in the granularity and frequency of data collected. For example, in vehicle tracking the data could be aggregated over larger time intervals providing some measure of trends which might offer reductions in premiums whilst ensuring greater privacy. Another idea would be to restrict the sharing of data until after an event had occurred. This would provide accurate data to assess the cause of the event. Privacy would be maintained whilst the insured individual would know that permanent monitoring was taking place. What is clear from these suggestions is that there are ways in which pervasive computing could be used more sensitively in a way which balances the demands of all.

This application of ICT is a far cry from the early days of data processing where the technology itself provided the constraint on the desire to collect, store and process every conceivable piece of data that might be useful. This technological constraint has gone. We must now look to our moral judgement to use technologies such as pervasive computing ethically. ICT professionals should share the burden with business professionals in defining acceptable use and establishing clear constraints. They must ensure that the subjects, for example, of insurance risk profiling, fully understand the terms and conditions they are agreeing to.

3.26 The Social Side of Wireless [2009]*

This section was first published as: Rogerson, S. (2009) ETHIcol—The social side of wireless. IMIS Journal, Vol 19 No 6, pp. 27–28. Copyright © Simon Rogerson.

Wireless technology has become an empowering technology for applications that have changed, challenged and sometimes confused many of us. As wireless began to be commonplace opinion was divided as to its impact. In 1994, Nicolas Negroponte wrote in *Being Digital*, "many of the values of a nation state will give way to those of both larger and smaller electronic communities. We will socialize in digital neighborhoods in which physical space will be irrelevant and time will play a different role." In contrast the following year, Clifford Stoll wrote in *Silicon Snake Oil*, "They isolate us from one another and cheapen the meaning of actual experience. They work against literacy and creativity. They will undercut our schools and libraries."

The evolutionary move to wireless has been marked first by miniaturisation in the 1990s and now functionality in the 2000s. Significant parts of our working and social lives are reliant upon wireless. We remain in touch on the move; we have an insatiable appetite for all kinds of information. Our wireless world is a world of anywhere, anytime, anyhow. The meaning of space and time has changed and this in turn has changed the way we behave.

There are currently 4.7 billion mobile connects globally and this has impacted upon our social behaviour. The mobile has transformed the space surrounding us from one of actual public space to one that is a virtual private space which moves with us. We can instantly disengage with the immediate physical space and engage in interactions of our choosing. As you read this article are you in a public space? Look around you—there will be many people in groups who are disengaged as they text and talk using their mobiles. Look at their companions. They will be adopting self-defence mechanisms such as drinking or reading as they feel socially isolated from their mobile-using friends. Of the people you can see, is there a difference in behaviour between young people and older people? This is because the young are able to co-exist in physical and virtual social settings. They communally text. They share communications openly. They will even swap mobiles so they can enter into new interactions. Older people tend not to behave like this.

Mobiles allow us to allocate time spent on interactions regardless of our physical location. We compress time spent on engaging in conversation and checking voicemails whilst travelling. We now treat these periods of time as units of resource which frees up quality time for something else. In some sense you are your mobile phone—could you turn yours off for an hour, a day or a week?

A wide range of wireless enabled applications once only available as separate instrument, such as PDAs, MP3 players and mobile phones, has now converged into single instruments such as the new iPhone 3G. The functionality is breathtaking—not only does it have all the existing functions of the separate instruments but there are many more. For example, it can be used to monitor your body signs during a workout and if you lose it you can locate it via satellite link and if it is not retrievable you can wipe the personal contents of the phone remotely. Today the fashion status of these devices is at least as important as the technical specification. The fashion trappings of the wireless world have entered the social psyche.

To sustain us we need water and energy. We treat these utilities in a special way. We expect them to be available on demand. Today our society as a whole and we as individuals need more and more information to sustain our activities. Increasingly we get this information via wireless. Wireless has become the new utility of the information age which we need to treat in the same special way as other utilities. This goes beyond the universal service obligation of reasonable access on an equitable basis regardless of location. Wireless should be the subject of new governance based on universal access and free at the point-of-service. This global utility model must be based on trust rather than economic imperative.

Wireless technology has changed our world. But we must remember we are all different. One solution is no solution because of the diversity of people due to age, culture, circumstance, and gender. For example, personal security is one of the most important reasons why people over

30 years of age have a mobile but this is not even and minor reason for young people. Young people also use their mobiles for entertainment and storing personal information but older people rarely use mobiles like this.

Wireless-enabled information availability must be sensitive to the social context of those who receive it. A recent piece of research considered how homeless people inhabit public spaces and suggested how wireless technology can create new opportunities of staying connected via synchronous rather than asynchronous communication thereby sustaining critical social networks to help them stay in touch because invisibility is a major danger for the homeless. Making accessible wireless communication available in these situations is challenging but one that must be overcome.

Social behaviour has and is changing with the advent of wireless technology. We have an obligation regarding technology roll-out to ensure that such changes are not detrimental to any of us and that we will all benefit. If we think of the wireless utility, then perhaps our approach will be different.

3.27 The Way to Healthy ICT? [2010]*

This section was first published as: Rogerson, S. & Haines, J. (2010) ETHIcol—The Way to Healthy ICT? IMIS Journal, Vol 20 No 1, pp. 35–36. Copyright © Simon Rogerson & Jemma Haines.

Personalised Health Monitoring (PHM) is an ICT growth area. PHM is concerned with using ICT to telemonitor individuals so, for example, chronic conditions are continuously checked remotely enabling healthcare to be administered at a distance, 24 hours, 7 days a week. PHM uses implantable, wearable, or portable ICT to collect vital body signs and biochemical statistics to determine the state of health of an individual. PHM can also contain information regarding activity, location, and the social and environmental contexts which can be important when judging state of health. The latest generation of PHM includes intelligent systems which combine monitoring data with biomedical data in undertaking automatic decision making.

Intel has been at the forefront of PHM. In their white paper *Emergence of Personal Health Systems* they state that PHM

> must go far beyond simply monitoring standard vital signs, or just adding new technical features. Instead, [PHM] must be able to address the needs of patients as people—connecting patients, caregivers, physicians, nurses, and others in an integrated, systematic, and interoperable way.

This will "provide both patients and healthcare professionals with real-time, interactive, data-rich health management systems that can engage both patients and their care management teams more fully in the treatment of their conditions."

So is PHM unqualified good news or do we need to be aware of any potential downsides? The product brief *Intel Health Guide PHS6000: connecting patients and healthcare professionals for personalised care* serves as an illustration. The brief explains that

> PHS6000 is a comprehensive personal health system that promotes greater patient engagement and more efficient care by combining an in-home patient device with an online interface, allowing clinicians to monitor patients and remotely manage care. It enables patients to: participate in their own care by monitoring their health status

under the guidance of a healthcare professional; communicate with healthcare professionals; and learn about their health and condition.

Now let us consider some elements of the product specification.

There is a strong emphasis on patient engagement using, for example video calls, to communicate with health care professionals. This is a laudable approach but the issue is the manner in which this might take place. PHS6000 has a one-way call system. If the daily medical statistics highlight a problem or if the patient has left a specific question to be answered then the clinician can call the patient. The patient cannot call the clinician. The system then rings like a phone and the patient can accept or reject the call. The system's web cam has a cover over it and can only be opened if the patient chooses to communicate visually as well as vocally.

This communication example raises some general issues. PHM interaction design needs to consider the role of synchronous or asynchronous communications, how the communication is instigated, what information is recorded and saved, and the nature of the communication interface. These issues are key to sustaining the relationship between patient and clinician, which is traditionally built upon trust in face-to-face interactions. The system's operational procedures are tailored for each patient. The system helps to monitor, and therefore prevent health deterioration and consequently helps to improve overall life quality. The way in which this is established can empower patients in new ways but could also impose new restrictive regimes on patients' lives. PHS6000 requires daily monitoring but is not time restrictive. Patients can monitor at any point during the day. Some conditions require fixed monitoring regimes whilst others do not and so perceived lost of freedom will differ. The selection of patients with particular conditions to use this type of PHM will certainly impact on the overall impression of PHM as an enabling or constraining facility.

Giving patients access to their health information is important; it can give them back control and enable successful health prevention monitoring. There are many impressive functions in PHS6000 which do this such as a general data reviewing function, a wide variety of activities which provide appropriate self-monitoring and a multimedia education library. There is one concern about this access. Medical information is a specialised area and the interpretation is complex. The language used can appear frightening to the layperson. The old adage *a little knowledge is a dangerous thing* is so true here. How a patient is provided with medical information in a way that helps rather than alarms has to be one of the biggest challenges for a system like PHS6000. This is further complicated because of the personal differences between clinicians who are using this system since these differences will generally lead to differences in service levels and patients' confidence of such service levels.

Like all medical systems PHM must address the privacy of patients. This is not simply an issue of legal compliance because current law may have shortcomings and indeed current law might be at odds with ethical demands. Two aspects of the PHS6000 specification need further investigation. The first concerns "results distributed to authorised professionals." It is unclear who is authorised and how they are authorised. Consideration of this issue also needs to be acknowledged by health care trusts implementing PHM devices. It is also unclear whether patients are informed of this distribution. The second issue concerns an implied attitude. The literature states that privacy is through encryption. The implication is that privacy has been turned into a technological issue rather than a social and ethical issue. If so then the addressing of privacy might be ethically limited.

Literature about the Intel Health Care management suite, of which PHS600 is a component, explains the suite "connects people and information in new ways which increase patient care and safety, reduce costs and improve quality of life." Such goals are worthy of support by society. PHS6000 has an "in home patient device online interface" developed using "patient centred design." But the question is who are the patients? Overall, there appears to be an assumption in the Intel literature of patients" technology literacy and no acknowledgement that a significant number of those who would benefit from this type of system may be nervous and often technology averse. Those using the system depicted in the literature appear to be from higher socioeconomic groups. Access by all to this advanced system through, for example using design-for-all and affordable point-of-use principles, needs to be established.

Further, issues to be considered regarding PHS600 do not just lie with the technology itself and the overall concept of PHM. Hypothetically if these were considered, debated, and then accepted serious consideration should then also occur for health care trusts/institutions accepting the use of such technologies. For example, is the organisation's infrastructure in place to meet the needs of PHM devices and ensure they are optimally delivered? Is the health economy able to meet the likely increased work demand when devices are introduced and rolled out? Who will be responsible for co-ordination of PHM within the workplace? Who will respond when devices fail and need repair? Who will train the patients in how to use the device, and then troubleshoot as required? Is there adequate additional clinician staffing to respond effectively to daily received data reports? Who is responsible for the duty of care when patients are using daily reporting mechanisms? Addressing such questions would be essential to ensure the devices would be implemented responsibly, as envisaged by its inventors and therefore facilitate it being fit for purpose.

Clearly PHS6000 offers much in advancing healthcare and it is recognised that any product brief will accentuate the positive. There does however need to be a response to the types of issues raised here in this brief analysis of the PHS6000. It is these types of issues that are the subject of the European FP7 research project PHM-Ethics which is analysing the relationship between ethics, law, psychosocial, medical sciences and ICT in the delivery of PHM. The project is indicative of much needed in-depth analysis of the application of ICT to healthcare which will ensure we all reap the massive potential benefits whilst ensuring the risks of such systems are minimised.

3.28 Ethical Robots [2011]*

*This section was first published as: ETHIcol—Ethical Robots. IMIS Journal, Vol 21 No 3, pp. 24–25. Copyright © Simon Rogerson.

On Wednesday 3 August 2011 the *Times* carried an article by Lucy Broadbent, "I, robot . . . you extinct?" It described how advances in robotics and artificial intelligence (AI) had led to the formation of The Singularity Institute which monitors such advances and whose members are top international scientists. There are biologists, biomedical gerontologists, cognitive scientists, cosmologists, philosophers, physicists, and software engineers within its membership.

The Institute claims that the computer is now much closer to being able to redesign itself and become more intelligent than humankind. This is the concept of singularity which the Institute explains as:

> The singularity is the technological creation of smarter-than-human intelligence. There are several technologies that are often mentioned as heading in this direction. The most commonly mentioned is probably "Artificial Intelligence," but there are others: direct brain-computer interfaces, biological augmentation of the brain, genetic engineering, ultra-high resolution scans of the brain followed by computer emulation. Some of these technologies seem likely to arrive much earlier than the others, but there are nonetheless

several independent technologies all heading in the direction of the singularity—several different technologies which, if they reached a threshold level of sophistication, would enable the creation of smarter-than-human intelligence.

(http://singinst.org/overview/whatisthesingularity/)

The purpose of the Institute is not to attempt to prevent AI advances, which is indeed an impossibility, but to ensure advances are benign and more likely to help humankind than damage it. This is a tall order, but one which is worthy of pursuit.

There is an interesting section concerning friendly AI on the website which includes this discussion of evolving moral values.

> The explicit moral values of human civilisation have changed over time and we regard this change as progress. We also expect that progress may continue in the future. An AI programmed with the explicit values of 1800 might now be fighting to re-establish slavery. Static moral values are clearly undesirable, but most random changes to values will be even less desirable. Every improvement is a change, but not every change is an improvement. Perhaps we could program the AI to "do what we would have told you to do if we knew everything you know" and "do what we would've told you to do if we thought as fast as you do and could consider many more possible lines of moral argument" and "do what we would tell you to do if we had your ability to reflect on and modify ourselves." In moral philosophy, this approach to moral progress is known as "reflective equilibrium."
>
> (http://singinst.org/summary)

There are many ethical issues surrounding the predicted advance to the Singularity. These issues are sometimes divided into two groups. Roboethics (Veruggio, 2005) focuses on the moral behaviour of humans in the design, use and treatment of AI beings whilst machine ethics (Anderson, 2011) is concerned with the moral behaviour of artificial moral agents such as robots. It is the latter which was the subject of a short article in *Scientific American* by Anderson and Anderson, Robot be good: a call for ethical autonomous machines published 14 October 2010.

Lin et al. (2011) bring together the concepts of singularity, artificial moral agents, roboethics, and machine ethics. The authors suggest there are three broad categories of issues: safety and errors, law and ethics, and social impact.

Concerning safety and errors, it is suggested that this is addressed during design and software development. However, if singularity is achieved it will not be an issue for humans, as is the case now, but an issue for an artificial agent. Questions the authors list under this category include:

- Is it possible for us to create machine intelligence that can make nuanced distinctions?
- What are the trade-offs between nonprogramming solutions for safety, such as restricted areas of robot deployment and the limitations they create?
- How safe ought robots be prior to their introduction into the marketplace or society?
- How would we balance the need to safeguard robots from running amok with the need to protect it from hacking or capture?
- How can we ensure robots only take salient, relevant safety information into account?

Concerning law and ethics the authors raise the issue of responsibility for action. As singularity is realised then it is reasonable that the robot has responsibility. Lin et al. explain that if robots "meet the necessary requirements to have rights, which ones should they have and how does one manage such portfolios of rights, which may be unevenly distributed given a range of biological and technological capabilities?" They raise a set of ethical questions that need to be answered and then embedded in law, which include:

- If we could program a code of ethics to regulate robotic behaviour, which ethical theory should we use?
- Are there unique legal or moral hazards in designing machines that can autonomously kill people?
- Is it ethically permissible to hand over responsibility for our elderly and children to machines?
- Will robotic companionship, such as drinking buddies, pets, other forms of entertainment, or sex, be morally problematic?
- Do we have any distinctive moral duties towards robots?
- At what point does a technology-mediated surveillance count as a "search," which would generally require a judicial warrant?
- Are there particular moral qualms with placing robots in positions of authority, such as police, prison or security guards, and teachers?

These sorts of questions are difficult to address but are further complicated by the fact that ethical and cultural norms and therefore law, vary around the world. No single existing legal jurisdiction is likely to be effective in the new world of singularity.

The final category of social impact focuses on the relationship between humankind and robots. Lin et al. pose the following questions:

- What is the predicted economic impact of robotics and how do we estimate the expected costs and benefits?
- Are some jobs too important or too dangerous for machines to take over?
- What do we do with the workers displaced by robots?
- How do we mitigate disruption to a society dependent on robotics, if those robots were to become inoperable or corrupted?
- Is there a danger with emotional attachments to robots and is there anything essential in human companionship and relationships that robots cannot replace?
- What is the environmental impact of a much larger robotics industry than we have today?
- Could we possibly face any truly cataclysmic consequences from the widespread adoption of social robotics and if so, should a precautionary principle apply?

With the emergence of more and more sophisticated AI we must address the moral issues surrounding it so we can ensure we still have some say in our destiny and that of future generations.

3.29 Ethics Omission Increases Gases Emission [2018]*

*This section was first published as: Rogerson, S. (2018) Ethics omission increases gases emission: A look in the rearview mirror at Volkswagen software engineering. Communications of the ACM, Vol 61 No 3, pp. 30–32, Copyright © the Association for Computing Machinery. Reprinted by permission. DOI DOI:10.1145/3180490

3.29.1 Introduction

The Volkswagen emissions scandal came to light in September 2015. The company installed software into millions of vehicles with diesel engines so that impressive emission readings would be recorded in laboratory conditions even though the reality is that the diesel engines do not comply with current emission regulations. Volkswagen is a worldwide organisation with its headquarters in Germany. Its subsidiaries adhere to common policies and a corporate culture. This worldwide scandal broke first in the US with ongoing investigation and legal action there and in other countries including Germany, Italy, and the UK.

Combustion engines are the source of pollution and therefore have been subjected to emission control. The formation of NOx (nitrogen oxides) through combustion is a significant contributor to ground-level ozone and fine particle pollution which is a health risk. On this basis, the use of software to control emissions must be defined as safety critical for, if it fails or malfunctions, it can cause death or serious injury to people. There does not appear to be any acknowledgement of this across vehicle manufacturing.

The statement from the US Department of Justice (2017) details the facts of the VW emissions case. Two senior managers, Jens Hadler and Richard Dorenkamp, appear to be at the centre of the so-called defeat software's ongoing design and implementation processes. These began in 2006, with the design of a new diesel engine to meet stricter US emission standards to take effect in 2007. The goal was to market new vehicles as meeting the stricter standards and attract US buyers. Being unable to accomplish this, the engineers working under Hadler and Dorenkamp, developed software which allowed vehicles to distinguish test mode from drive mode thus satisfying the emissions test whilst allowing much greater emissions when vehicles were on the road. "Hadler authorized Dorenkamp to proceed with the project knowing that only the use of the defeat device software would enable VW diesel vehicles to pass U.S. emissions tests."

Drawing upon *the Statement of Facts*, Leggett (2017) reported that whilst there had been some concerns over the propriety of the defeat software all those involved in the discussions including engineers were instructed not to get caught and furthermore to destroy related documents. According to Mansouri (2016), Volkswagen is an autocratic company with a reputation for avoiding dissent and discussion. It has a compliant business culture where employees are aware that underperformance can result in replacement and so management demands must be met to ensure job security. Three statements in particular in the *Volkswagen Group Code of Conduct* (Volkswagen, 2010): *Promotion of Interests* (ibid, p. 15), *Secrecy* (ibid, p. 16) and *Responsibility for Compliance* (ibid, p. 22), align with the ongoing conduct encouraged during the emissions debacle. Trope and Ressler (2016) explain that as an autocratic book of rules, the group code supports and even promotes dishonest dysfunctional behaviour which includes the creation of software to cheat, rather than solve, engineering problems and to protect that software from disclosure as if it were a trade secret.

On 11 January 2017, the US Justice Department announced that,

> Volkswagen had agreed to plead guilty to three criminal felony counts, and pay a $2.8 billion criminal penalty, as a result of the company's long-running scheme to sell approximately 590,000 diesel vehicles in the U.S. by using a defeat device to cheat on emissions tests mandated by the Environmental Protection Agency (EPA) and the California Air Resources Board (CARB), and lying and obstructing justice to further the scheme.

3.29.2 Business Analysis

Many of the accounts about the Volkswagen emissions case focus on business ethics with only a few touching upon the role of the software engineers in this situation. These accounts at times are repetitive but intertwine to provide a rich view. The widespread unethical actions across Volkswagen can be described as a new type of irresponsible behaviour, namely *deceptive manipulation* (Siano et al., 2017). The detail of this and the associated corporate repercussions are discussed further by Stanwick and Stanwick (2017).

Software engineers at Volkswagen faced ethical and legal issues that are easy to identify. Plant (2015) suggests that they should have alerted external bodies since the internal lines of reporting were compromised. Merkel (2015) concurs citing the *Software Engineering Code of Ethics and Professional Practice* (see www.acm.org/about/se-code) by way of justification, and adds that the lack of whistleblowers in such a large group is surprising. Both authors point to the potential personal cost of whistleblowing as the reason it did not happen. Rhodes (2016) adds a second factor, arguing that corporate business ethics is very much a *pro-business stance* which is implemented through corporate control and compliance systems, and instruments of managerial coordination. This can enable the pursuit of business self-interest through organised widespread conspiracies involving lying, cheating, fraud and lawlessness. This is what happened at Volkswagen. Queen (2015) concurs, explaining that Volkswagen intentionally deceived those to whom it owed a duty of honesty. The pressure for continuous growth and the perception that failure was not an option (Ragatz, 2015) created a culture where corporate secrecy was paramount which in turn implicitly outlawed whistleblowing.

3.29.3 The Role of Software Engineering

If one has a responsibility for the planning, design, programming, or implementation of software then that aspect of one's work falls within the scope of the *Software Engineering Code of Ethics and Professional Practice* regardless of one's job title. In that sense software engineering pervades this debacle and is therefore worthy of further investigation.

So what was the role of software engineers in the creation and installation of VW's defeat software? This question can be addressed using the *Software Engineering Code of Ethics and Professional Practice*. The code is long established, documenting the ethical and professional obligations of software engineers and identifying the standards society expects of them (Gotterbarn et al., 1999). The code translates ethical principles into practical guidance. It encourages positive action and resistance to act unethically. It has been adopted by many professional bodies and companies worldwide and has been translated into Arabic, Croatian, French, German, Hebrew, Italian, Mandarin, Japanese, and Spanish.

Software engineers and software engineering educators have a responsibility to be cognisant of the code and its requirements. *Public Interest* is central to the code which is apposite for safety critical software. Although education can influence the courage and capability to act in accordance with the code, that result depends on structural and psychological supports within the environment in which engineers practice.

The actions of VW managers and software engineers violated the following principles of the code:

Principle 1.03 "approve software only if they have a well-founded belief that it is safe, meets specifications, passes appropriate tests, and does not diminish quality of life, diminish privacy, or harm the environment. The ultimate effect of the work should be to the public good." The defeat software is clearly unsafe given NOx pollution damages both health and the environment. The public were under the misapprehension that VW cars were emitting low levels of NOx and therefore not a health risk. Thus, software engineers installed unethical software.

Principle 1.04 "disclose to appropriate persons or authorities any actual or potential danger to the user, the public, or the environment, that they reasonably believe to be associated with software or related documents." There is no evidence that any software engineer disclosed. Commercial software is usually developed in teams and in this case it is likely this was a large team spanning all aspect of software development.

Principle 1.06 "be fair and avoid deception in all statements, particularly public ones, concerning software or related documents, methods and tools." The emissions software was heralded publicly as a success when internally there was widespread knowledge that this claim was fraudulent. Software engineers were likely to have been privy to this cover-up.

Principle 2.07 "identify, document, and report significant issues of social concern, of which they are aware, in software or related documents, to the employer or the client." There is some evidence that concern was raised about the efficacy of the defeat software but it seems those in dissent allowed themselves to be managed towards deception.

Principle 3.03 "identify, define and address ethical, economic, cultural, legal and environmental issues related to work projects." The EPA regulations are explicit and are legally binding. From the evidence accessed it is unclear as to whether software engineers knew of the illegality of their actions. Nevertheless, ignorance cannot and must not be a form of defence.

Principle 6.06 "obey all laws governing [the] work, unless, in exceptional circumstances, such compliance is inconsistent with the public interest." This relates to the analysis under principle 3.03. Compliance to further the prosperity of Volkswagen was at the expense of legal compliance.

Principle 6.07 "be accurate in stating the characteristics of software on which they work, avoiding not only false claims but also claims that might reasonably be supposed to be speculative, vacuous, deceptive, misleading, or doubtful." Software engineers could argue internally that the software indeed performed as it was designed to. However, the design was to achieve regulatory and public deception.

Principle 6.13 "report significant violations of this Code to appropriate authorities when it is clear that consultation with people involved in these significant violations is impossible, counter-productive or dangerous." Given the apparent corporate culture within Volkswagen there was little point in reporting concerns further up the line. In fact, the corporate code seems at odds with the professional code regarding this point. Software engineers failed to report these breaches to appropriate authorities.

3.29.4 Conclusions

Professionals, who must have been party to this illegal and unethical act, developed and implemented this software. Those who undertake the planning, development, and operation of software have obligations to ensure integrity of output and overall to contribute to the public good (Rogerson, 2011). The ethical practice of software engineers is paramount. Practice comprises process and product. Process concerns virtuous conduct of software engineers, whereas product concerns whether software is deemed to be ethically viable. Actions and outcomes in the Volkswagen case appear to have failed on both counts.

These serious issues related to professional practice need to be addressed. It is hoped such issues are exceptional but sadly it is likely they are commonplace given the ongoing plethora of software disasters (see, for example, Catalogue of Catastrophe (2016) and Software Fail Watch (2016)). Unethical actions related to software engineering can be addressed from two sides. One side focuses on resisting the temptation to perform unethical practice whilst the other focuses on reducing the opportunity of performing unethical practice. Society at large needs competent, ethical, and altruistic professionals to deliver societally-acceptable, fit-for-purpose software. Both of these can be helped by education, but education will not suffice without adequate social supports.

In order to fulfil software engineering duties, an individual must fully understand the professional responsibilities and obligations of the role. These are explicitly laid out in the *Software Engineering Code of Ethics and Professional Practice* and as such individuals must know and apply it to their everyday work. To achieve this, the effective education of new professionals is essential. Teaching technology in isolation is unacceptable and dangerous. Software engineers need a broader education to gain the necessary skills and knowledge to act in a socially responsible manner not on the basis of instinct and anecdote but on rigour and justification (Rogerson, 2015). They must possess practical skills to address the complex ethical and societal issues which surround evolving and emerging technology. Such education should be based on a varied diet of participative experiential learning delivered by those who have a practical understanding of the design, development, and delivery of software. Contrasting the Volkswagen Group Code of Conduct with the Software Engineering Code might provide one means for experiential learning. Such educated software engineers might find ways to prevent the installation of unethical software of the future.

3.30 People are Data in the Connected World [2018]*

This section is a slightly modified version of an article first published as: Rogerson, S. (2018) The connected world and mobility: Ethical challenges. Internet of Business, 14 September. Additional reporting by Chris Middleton, http://internetofbusiness.com/the-data-self-the-connected-world-and-mobility-a-global-ethical-challenge/. Copyright © Simon Rogerson. A second version of this appeared as Rogerson, S. (2018) Wireless and Social Media Influence. International Journal for the Data Protection Officer, Privacy Officer and Privacy Counsel, Vol 2 No 7, pp. 12–16. Copyright © Simon Rogerson.

Wireless technology has transformed the space surrounding us from a physical public space into a virtual private one that moves everywhere with us.

We can instantly disengage with the physical world and engage in virtual interactions of our choosing—on what we perceive to be a global basis. The smartphone revolution was one of the final pieces of this jigsaw, but the Internet of Things (IoT) will complete the picture, providing a virtual shadow of the physical world.

This we all know. But the very meaning of space and time has changed and this, in turn, has changed the way we behave in this new space. The ethical and social implications of that cannot be underestimated.

We've become digital citizens with our needs satisfied and aspirations fulfilled through a combination of the real and virtual worlds—as the people surrounding you in cafes, on trains, and in physical public spaces attest, as they gaze into their screens.

The speed of technological advance is astonishing—as illustrated by the following statistics. In 2010, 400 million people had Facebook accounts, 126 million blogs existed, 50 million tweets were created daily, and 91% of mobile Web users accessed social networking sites.

But by 2018, that picture has changed dramatically: 2.2 billion people now have Facebook accounts; over 440 million blogs exist on Tumblr, Squarespace, and WordPress alone; more than five billion videos are watched every day on YouTube; 500 million tweets are created daily; 546 million people are LinkedIn users in over 200 countries; and most online/social time is spent on smartphones, with WhatsApp boasting 1.5 billion active users.

Social media, supported by high-performance wireless, is now an essential part of the global societal infrastructure. Its variety and reach continue to expand. Consultant and tech blogger Fred Cavazza's diagram of this expanding landscape clearly shows that many of the activities, including networking, publishing, sharing, messaging, discussing, and collaborating, that provide the "social glue" of the human world are increasingly supported through social media. This raises a number of ethical issues, such as access, technological literacy and aversion, and economic disparity.

An analysis of internet penetration undertaken by Hootsuite in January highlights these issues. It's clear from this analysis that regions of socio-economic deprivation, such as central Africa, are increasingly disadvantaged in a world dependent on online services. Put another way, many of these people are not currently part of the same conversation.

3.30.1 Privacy

Privacy is another critical area for ethics in the connected age. In 2011, Viviane Reding, vice president of the European Commission and EU Justice commissioner, laid out the foundations on which new data protection regulations should be based.

These comprised four pillars: the right to be forgotten; transparency; privacy by default; and protection regardless of data location. In May 2018, the European General Data Protection Regulation (GDPR), which is the third version of European data protection legislation, was rolled out.

It contains a list of an individual's rights as to how personal data is handled. These include the rights: to be informed; of access; to rectification; to erasure; to restrict processing; to data portability; to object; and not to be subject to automated decision-making, including profiling.

The US is now on the cusp of adopting similar rules, some believe.

California has a history of being in the vanguard of privacy legislation. In 1972, voters amended the state's Constitution to include the legal and enforceable right to privacy as being among the "inalienable" rights of all citizens. However, over the past quarter century, that right has been encroached on by the digital economy—ironically, led by companies in the state, such as Google and Facebook.

In November 2017, lawyers acting on behalf of the citizens of California wrote to the Attorney General, outlining proposals for a new consumer privacy act. Their proposed law entailed adding 15 clauses to the state's Civil Code. The most significant ones for data-collecting organisations such as social platforms, were:

- the right to know what personal information is being collected
- the right to know if personal information is sold or disclosed, and to whom
- the right to say no to the sale of that personal information
- the right to equal service and price (i.e., not to be discriminated against, based on that personal data)

More, the draft legislation's definition of personal information was extremely broad, and included:

- identifiers such as name, address, IP address, email address, account name, social security number, passport number, and driving licence
- property records
- biometric data
- browsing history, interaction with advertisements, apps, or websites
- geolocation data
- audio, electronic, visual, thermal, olfactory, or similar information, including facial recognition
- psychometric data
- employment history
- inferences drawn from any of the information identified previously
- all of the above as applied to any minor children of the data subject

On 28 June 2018, the California Consumer Privacy Act of 2018 (CCPA) was passed unanimously. However, it watered down some of the November proposals.

Most significantly, it includes an exception to the right to equal service, allowing companies to offer different levels of service, depending on how customers interact with a site, app, or advertisement—the so-called "Spotify exception."

3.30.2 Corporate Support

Google, Facebook, and others, oppose tighter regulations of any kind, which is why they've been lobbying the US government for a watered-down federal solution that serves their commercial interests. The aim is to neuter California's act.

But Europe is getting involved once again. In September 2018, France's data regulator, the Commission Nationale de l'Informatique et des Libertés, announced that it is seeking to extend the so-called right to be forgotten globally, arguing that any Europe-only removal of data is meaningless on a global platform in an age of IP cloaking.

Either way, new European and US state legislation acknowledges something critically important: that personal data is now *an important part of an individual* and, consequently, the individual must have much greater control over that data, given that it is now a fundamental element of digital citizenship.

Acknowledging that humans are becoming composite beings, in effect, leads to a requirement to think of ourselves not as data *subjects*, but as *data selves*.

It therefore follows that, data—our virtual anatomy, if you like—should never be owned by third parties. This additional right of the individual will perhaps be included in the fourth version of data protection legislation sometime in the future.

The problem is that there is currently no widely accepted mechanism for managing the digital self online, such as a personal API, although some blockchain projects are working in this area.

3.30.3 Not the Only Ethical Issue

While privacy is an important ethical issue in the context of social media, it is by no means the only one as the connected world expands, along with its ethical challenges.

Moral norms and values are embedded in social networking sites. As such, according to Light and McGrath (2010), the technology shapes the user experience and, ultimately, changes the user as more and more time is spent on a platform.

Bateman et al. (2011) explain that self-disclosure on social media has begun to change what spaces, time, and information we judge to be private or public. This tension raises a number of ethical issues.

Dual use and even multiple uses are commonplace across social media. Witt (2009) provides evidence of employers perusing social profiles before making hiring decisions. Lawyers often access social media to collect incriminating information in divorce and child custody cases, she says.

Thus, social platforms present a danger to participants and society as a whole, if improperly accessed or used.

Social network sites can be, and are, monitored, and legal precedents have been established. For example, Strutin (2011) explains how a juvenile gang member pleaded guilty to a weapons offence in a California court. He was sentenced to probation, which included barring him from access to any social network.

This judicial recognition of social networks as communication media that can be monitored should change our perception of them.

Meanwhile, Louch et al. (2010) suggest that the teenage years are when adolescents work on determining their identity. Advertisements, which are core to the social media experience, have an impact on their sense of who or what they can become.

As such, it could be argued that targeted marketing using social profiles is tantamount to covert social manipulation, which is discriminatory. Such action requires ethical scrutiny.

This challenging ethical landscape was amply illustrated by the recent problems with Facebook and Cambridge Analytica, a scandal in which 87 million data profiles were shared.

Blogging on Elastic Creative on 23 March 2018, Melrose explains that data selves were harvested by Cambridge Analytica, using a digital personality quiz, together with data from their Facebook profiles and information about their friends.

It was "people as data" who were the priceless commodities mined by Cambridge Analytica, and subsequently used for a disingenuous purpose.

This type of unethical behaviour is not uncommon. On 21 July 2018, the BBC reported that Facebook had suspended Crimson Hexagon, a US-based analytics firm, while it investigated concerns about the collection and sharing of user data.

3.30.4 To Conclude

The ethical issues surrounding social media are complex and difficult to address, and consideration of them must go beyond mere compliance with current regulations and laws.

Creating simple checklists is problematic, as the ethical dimension of ICT can, and does, change rapidly as technology evolves. The appliance of ethical sensitivity—rather than compliance with an ethical checklist—must be the way forward.

Our data shadows will remain for as long as the virtual world exists. We are, therefore, all permanently vulnerable in this technologically dependent world.

3.31 Hospital Safety [2021]*

*This section is based on my involvement with the SAFECARE project as a member of its Ethics Board. The description of the project is drawn from published material posted on www.

safecare-project.eu and cyberwatching.eu/projects/2688/safecare/news-events/safecare-creating-integrated-cyber-physical-security-system-healthcare-infrastructure. The material on SAFECARE ethics was my response to a request for advice on how to address ethics in post-implementation situations. Copyright © Simon Rogerson.

The lines between the physical and cyber world are becoming increasingly blurred as the Internet of Things takes off and digital connections become ubiquitous. In areas where this is not currently the case, physical intrusion may break down barriers. Threats can no longer be analysed solely as physical or cyber and it is therefore critical to develop an integrated approach in order to fight against such combination of threats. Health services are among the most critical infrastructures, and most vulnerable.

SAFECARE is a research and development project funded by the European Union's H2020 research and innovation programme. The aim of SAFECARE is to provide solutions that will improve physical and cyber security in a seamless and cost-effective way. Thereby, it promotes new technologies and novel approaches to enhance threat prevention, threat detection, incident response, and mitigation of impacts. The project has conducted research in four areas:

- healthcare infrastructure threat assessment and solution requirements
- physical security solutions for healthcare infrastructure
- cybersecurity solutions for healthcare infrastructure
- integrated Cyber-Physical security solutions for healthcare infrastructure

SAFECARE has created an integrated cyber-physical security system for healthcare infrastructure. In basic terms, this means that it is taking inputs from two separate systems (physical and cyber) and displaying them through one interface panel and security software. The innovative part of the project is building an integrated system so that it analyses separate incidents from both systems and classes them as innocent or elevates them to a threat that a member of the security team should consider. The SAFECARE prototype has been successfully demonstrated in hospitals in Turin and Marseille.

SAFECARE provides a holistic approach to address not only cyber but also integrated cyber-physical threats. The innovative and cost-effective cyber-physical security tools offered enhanced prevention, detection, response, and mitigation capacities, integrated with hospital systems and without interference to the medical services, doctors tasks and patients' care. Critical assets, vulnerabilities, threats, and risks, specifically of the healthcare sector, are analysed leading to real and focused cyber situation awareness. SAFECARE will improve end-user awareness by providing them structured crisis management and an enlarged and relevant vision during the response, leading to enhanced resilience.

During the SAFECARE project the surrounding ethical issues have been scrutinised so that the project's deliverables have a reasonable chance of being ethically fit-for-purpose in operational situations. This has been challenging given the hybrid nature of the operating environment. The three possibilities of attack are: cyber; physical; and a combination of cyber and physical. The latter is further complicated as the ratio of combination can vary considerably.

SAFECARE Ethics is encapsulated in the project name. All those working in the health sector, regardless of their role, have a duty of care so that patients and their associates feel safe. Therefore, in operational situations, the public must be able to trust those working with SAFECARE and must have confidence that the system, with its many complex interrelated components, is trustworthy. The visibility of ethical consideration is important to ensure trust and trustworthiness. This must be the focus of post-implementation SAFECARE Ethics. There are two distinct ethical dimensions to consider:

process and *product*. In this situation, process concerns the activities of those providing SAFECARE service/product delivery; the key areas to focus on being education and training, governance, and professional conduct. In this situation, product concerns the potential impact of SAFECARE on people and society; the key areas to focus on being embedded ethical values and technological integrity. When SAFECARE becomes operational there must be effective training of all associated staff to ensure full commitment to SAFECARE Ethics rather than superficial compliance.

Here are some of the key ethical issues which relate to SAFECARE and whether these are considered as issues of process or product or both.

- The creation of false positives and false negatives. For example, in video surveillance, appliances, human movement and apparel may be wrongly identified by AI (Product).
- Overriding system decisions by human intervention (Process and Product).
- Ongoing monitoring and evaluation of system and operative performance particularly from an ethical perspective (Process).
- Legitimate dual use of data in situations where there are several ethically justified actions. For example, video surveillance data could lead to patient and staff safety but may also be problematic due to privacy (Product).
- SAFECARE transparency to ensure public awareness, understanding and acceptance (Process and Product).
- Pragmatic ethics training of all those associated with any aspect of SAFECARE operation (Process).
- Consideration of the severity of an identified incident and its impact on immediate patient wellbeing and staff safety (Process and Product).
- Identification, due consideration and involvement of all stakeholders who are directly or indirectly affected by SAFECARE. For example, this should include families of patients (indirect) as well as patients (direct) (Process).
- There are obvious ethical issues such as privacy but there will also be less obvious ones and some yet to be identified. The approach to SAFECARE Ethics should take this into account (Process and Product).

There is a need to develop a vision for SAFECARE Ethics which is underpinned by theory but practically viable, so that all will engage, accept, and embrace this vision as a modus operandi. Any barriers between physical security functions and cyber security functions must be overcome as healthcare safety is reliant upon both functions working in harmony thus providing a synergistic deterrent to any form of attack. Thus, the SAFECARE approach has the potential to spawn a new inclusive co-operative culture within and across hospitals.

References

Anderson, S. L. (2011). Machine metaethics. In M. Anderson & S. L. Anderson (Eds.), *Machine ethics* (pp. 21–27). Cambridge: Cambridge University Press.

Bateman, P. J., Pike, J. C., & Butler, B. S. (2011). To disclose or not: Publicness in social networking sites. *Information Technology & People, 24*(1), 78–100.

Catalogue of Catastrophe (2016). *International project leadership academy*. Retrieved March 29, 2017, from http://calleam.com/WTPF/?page_id=3.

Gotterbarn, D., Miller, K., & Rogerson, S. (1999, October). Software engineering code of ethics is approved. *Communications of the ACM, 42*(10), 102–107 and *Computer. Oct*, 84–89.

Leggett, T. (2017, January 12). VW papers shed light on emissions scandal. *BBC News*. Retrieved January 17, 2017, from www.bbc.co.uk/news/business-38603723.

Liao, G.-Y., & Hwang, J.-J. (2001). A trustworthy internet auction model with verifiable fairness. *Internet Research: Electronic Networking Applications and Policy*, *11*(2), 159–166.

Light, B., & McGrath, K. (2010). Ethics and social networking sites: A disclosive analysis of Facebook. *Information Technology & People*, *23*(4), 290–311.

Lin, P., Abney, K., & Bekey, G. (2011). Robot ethics: Mapping the issues for a mechanized world. *Artificial Intelligence*, *175*(5–6), 942–949.

Louch, M. O., Mainier, M. J., & Frketich, D. D. (2010). An analysis of the ethics of data warehousing in the context of social networking applications and adolescents. *2010 ISECON Proceedings*, *27*(1392).

Mansouri, N. (2016). A case study of Volkswagen unethical practice in diesel emission test. *International Journal of Science and Engineering Applications*, *5*(4), 211–216.

Merkel, R. (2015, September 30). Where were the whistleblowers in the Volkswagen emissions scandal? *The Conversation*. Retrieved September 14, 2016, from http://theconversation.com/where-were-the-whistleblowers-in-the-volkswagen-emissions-scandal-48249.

Plant, R. (2015, October 15). A software engineer reflects on the VW scandal. *The Wall Street Journal*. Retrieved January 15, 2017, from http://blogs.wsj.com/experts/2015/10/16/a-software-engineer-reflects-on-the-vw-scandal/.

Ponemon, L. (2004). *Consumer survey on identity management*. Ponemon Institute.

Porter, M. (1985). *Competitive advantage – Creating and sustaining superior performance*. The Free Press.

Queen, E. L. (2015, September 26). How could VW be so dumb? Blame the unethical culture endemic in business. *The Conversation*. Retrieved September 15, 2016, from http://theconversation.com/how-could-vw-be-so-dumb-blame-the-unethical-culture-endemic-in-business-48137.

Ragatz, J. A. (2015). What can we learn from the Volkswagen scandal? *Faculty Publications*. *Paper 297*. Retrieved September 6, 2016, from http://digitalcommons.theamericancollege.edu/faculty/297.

Rhodes, C. (2016). Democratic business ethics: Volkswagen's emissions scandal and the disruption of corporate sovereignty. *Organization Studies*, *37*(10), 1501–1518.

Rogerson, S. (2011). Ethics and ICT. In R. Galliers & W. Currie (Eds.), *The Oxford handbook on management information systems: Critical perspectives and new directions* (pp. 601–622). Oxford University Press.

Rogerson, S. (2015). Future vision. *Special Issue—20 years of ETHICOMP, Journal of Information, Communication and Ethics in Society*, *13*(3/4), 346–360.

Siano, A., Vollero, A., Conte, F., & Amabile, S. (2017). "More than words": Expanding the taxonomy of greenwashing after the Volkswagen scandal. *Journal of Business Research*, *71*(C), 27–37.

Software Fail Watch (2016). Quarter one. *Tricentis*. Retrieved March 29, 2017, from www.tricentis.com/blog/2016/04/07/software-fail-watch-2016-quarter-one/.

Stanwick, P., & Stanwick, S. (2017). Volkswagen emissions scandal: The perils of installing illegal software. *International Review of Management and Business Research*, *6*(1), 18.

Strutin, K, (2011). Social media and the vanishing points of ethical and constitutional boundaries. *Pace Law Review*, *31*(1), article 6.

Taylor, F. W. (1911). *The principles of scientific management*. Harper & Brothers.

Trope, R. L., & Ressler, E. K. (2016). Mettle fatigue: VW's single-point-of-failure ethics. *IEEE Security & Privacy*, *14*(1), 12–30.

US Department of Justice (2017, January 11). Volkswagen AG agrees to plead guilty and pay $4.3 billion in criminal and civil penalties and six Volkswagen executives and employees are indicted in connection with conspiracy to cheat U.S. emissions tests. *Justice News*. Retrieved January 15, 2017, from www.justice.gov/opa/pr/volkswagen-ag-agrees-plead-guilty-and-pay-43-billion-criminal-and-civil-penalties-six.

Veruggio, G. (2005). *The birth of roboethics*. ICRA 2005, IEEE International Conference on Robotics and Automation, Workshop on Roboethics.

Volkswagen (2010). *The Volkswagen group code of conduct*. Retrieved January 28, 2017, from http://en.volkswagen.com/content/medialib/vwd4/de/Volkswagen/Nachhaltigkeit/service/download/corporate_governance/Code_of_Conduct/_jcr_content/renditions/rendition.file/the-volkswagen-group-code-of-conduct.pdf

Witt, C. L. (2009). Social networking: Ethics and etiquette. *Advances in Neonatal Care*, *9*(6), 257–258.

Chapter 4

Practice and Method

This is the first of three chapters which cover implementation. In the context of this book, practice is a term which usually refers to the conduct as well as the work of someone from a specific profession (for example, see Section 4.3) whereas method is a process by which a task is completed (for example, see Section 4.18). Kemmis (2009) has concerns about the commonly-accepted meaning of professional practice. He explains (ibid, p. 160) that professional "practices have historical and social consequences, meaning and significance for communities and societies which are beyond the particular measurable effects (and effectiveness) of particular acts of particular practitioners at particular times." Kanes (2009) reports that the multiple perspectives which Kemmis puts forward suggests there should be a balance between the subjective and objective views of practice and that this would counter ongoing technological rationality. This is important because technological rationality is totalitarian, resulting in technology and industry controlling the structure of the economy, intellectual pursuits, and leisure activities (Marcuse, 1964).

In many ways the accounts within this book are evidence that Marcuse's concerns over technological rationality are well founded. Codes of conduct are the traditional instruments of guidance for acceptable professional practice (see Section 5.6). However, there are other approaches to promoting good practice. There are two aspects to realising ethical digital technology development and use: to resist the temptation to perform unethical practice and to reduce the opportunity of performing unethical practice. These are addressed either reactively, where the measure is in response to a particular event or circumstance, or proactively, where the measure is an attempt to promote desired future behaviour. The latter is more important, as it will lead to lasting positive change. Some examples of proactive actions are (Rogerson, 2021, p. 193):

- resist temptation
  - education and Training programmes
  - mandatory ethics committees associated with operational actions
- reduce opportunity
  - regulation and policies frameworks (see Chapter 5)
  - public awareness programmes which generate public pressure

Suggestions can be found in many of the sections in this chapter as most sections discuss the challenges arising from a particular digital technology application or incident. These discussions

DOI: 10.1201/9781003309079-4

include possible changes in practice or method to ensure that the particular ethical problem is unlikely either to go unnoticed or to reoccur in the future.

4.1 Training for Ethics [1996]*

This section was first published as: Rogerson, S. (1996) ETHIcol—Training for Ethics. IDPM Journal, Vol 6 No 1, p. 26. Copyright © Simon Rogerson.

Society in general and organisations in particular are becoming more dependent upon computer technology. Those responsible for the development and application of computer technology are faced with decisions of increasing complexity which are accompanied by many ethical dilemmas. Computer technology is a special and unique technology, and hence the associated ethical issues warrant special attention. Indeed, there is an urgent need to understand the basic cultural, social, legal, and ethical issues inherent in the computing. For these reasons it is imperative that future computer professionals are taught the meaning of responsible conduct so that they do not fall into the bad habits of their predecessors. The inclusion of computer ethics in a professional development programme will develop in individuals a reasoned and principled process by which reflective moral judgements can be made.

It is important to be very clear as to what is being addressed by the inclusion of computer ethics for the use of the term "ethical dilemmas" can be misleading. A description of the concept of a dilemma in the Encyclopaedia Britannica includes a warning. It is pointed out that it is not necessary that a dilemma should have an unwelcome conclusion; but from the use in rhetoric the word has come to mean a situation in which each of the alternative courses of action (presented as the only ones open) leads to some unsatisfactory consequence. This misconception must be countered by a precise definition of the professional development programme. Consequently, the pedagogical objectives of the professional development programme within computing should include:

- providing an introduction to the responsibilities of the profession
- articulating the standards and methods used to resolve non-technical ethics questions about the profession
- developing some proactive skills to reduce the likelihood of future ethical problems

These objectives satisfy the second level of computer ethics, termed "para" computer ethics. Para computer ethics is concerned with identifying, clarifying, comparing, and contrasting computer ethics cases. This results in practical resolutions of ethical dilemmas.

All computer professionals and policy makers should have these second level skills and knowledge in order to do their jobs effectively and sensitively.

Formal training whilst including some conceptual underpinning should focus primarily upon experiential learning through the use of case studies. The intention is, firstly, to educate the participant in the ethical principles which should underpin the work of the computer professional and, secondly, to train the participant in ethical case analysis. There are eight ethical principles that the computer professional should subscribe to. These relate to honour, honesty, bias, adequacy, due care, fairness, social cost, and action. All should be addressed in the programme. Turning to case analysis, there is a need to provide a framework which enhances critical analysis and overcomes shortcomings in the understanding of ethical theories. This process decomposes the complex problem into its constituent parts which in turn enhances understanding and increases the chances of deriving a solution that is reasoned out and ethically sensitive.

Whilst formal training is important as it provides the essential foundations in computer ethics it will be worthless without some form of ongoing development and support. Newly acquired enlightened attitudes need to be nurtured. Acting from an ethical standpoint can often lead to conflict both with peers and corporate objectives. Support and encouragement together with a wider ownership of ethically sensitive decisions and actions must be achieved. There are several commonly accepted work practices that could be adopted to affect this change. For example, consider the interesting parallels between quality assurance and applied ethics. In both cases, there is a need to consider the related issues throughout the process and to encourage work practices and attitudes which promote the associated fundamental concepts. At the same time, these "softer" concepts are often misconceived, by both management and the workforce, as being secondary to "harder" concepts such as finance. In the late 1970s and throughout 1980s, quality became more central to organisation strategy. With the realisation that quality was a prerequisite to economic security many organisations in the western industrialised countries, notably the US and the UK, turned towards Japan to gain new insights into quality assurance. Some of the initiatives succeeded while others did not. One of the most interesting concepts to be adopted was the Quality Circle. This is a small group of people who do similar work meeting voluntarily on a regular basis to identify problems, analyse the causes, recommend the solutions to management and, where possible, implement those solutions themselves. Given the parallels between quality and ethics, the establishment of Ethics Circles to support the consideration of ethical issues in organisations has merit. These could be operated on similar lines to Quality Circles thus providing support to individuals who are engaged in ethically sensitive decision making, raising general awareness of ethical issues, and acting as informal staff development in this area.

The combination of formal training through, for example, case study analysis and informal support through, for example, Ethics Circles should ensure that computer professionals within organisations are equipped to handle the ethical dilemmas which they will inevitably face daily. Each day the technology invades another small part of our lives. Each day we become a little more dependent upon the technology. Computer professionals must be trained so that they are sensitive to the power of the technology and act in a responsible and accountable manner. Only then will there be a chance for the realisation of an empowering technology rather than an enslaving one.

4.2 Social Responsibility [1996]*

*This section was first published as: Rogerson, S. (1996) ETHIcol—Social Responsibility. IDPM Journal, Vol 6 No 3, p. 17. Copyright © Simon Rogerson.

In previous editions of this column many fundamental issues have been raised concerning the sensitive use of Information Technology and the responsibilities computer professionals have as custodians of IT. Society at large is becoming more aware of the fact that IT can empower or enslave and the choice made quite often is reliant upon the professional attitudes of the technologists. It is concerning therefore that the practitioners within the profession are somewhat complacent about these very important issues. It is very concerning and disappointing that not a single practitioner has ever responded to the issues raised in this column. Why is this so? Exploitation of IT in the name of commercial economics or so-called commercial reality is no excuse. Those reading this column who develop and implement systems should remember that the first question should always be, "Is it ethical?"

In a recent article titled "Unique Ethical Problems in Information Technology" Walter Maner (1996) discusses why special attention should be paid to the issues and problems unique to the field of computer ethics. He explains uniqueness refers to those ethical issues and problems that:

- are characterised by the primary and essential involvement of IT
- exploit some unique property of that technology
- would not have arisen without the essential involvement of IT

The facets of this uniqueness include storage, complexity, adaptability and versatility, processing speed, relative cheapness, reproduction, and coding. This uniqueness has resulted in a failure to find satisfactory non-computer analogies that might help us to resolve IT-related ethical dilemmas. According to Maner, this "forces us to discover new moral values, formulate new moral principles, develop new policies, and find new ways to think about the issues presented to us."

One of the many examples Maner uses to illustrate his argument concerns the superhuman complexity of computers. We might program computers but it is commonplace that the performance of such programs defies inspection and understanding. It might be possible to understand program code in its static form but it does not follow that we understand how the program works when it executes. We have come to rely heavily upon IT and we are continually attempting to make it perform to its limits and often beyond because we do not fully understand its complex nature.

In the UK, for example, Nuclear Electric decided to rely heavily on computers as its primary protection system for its first nuclear power plant, Sizewell B. The company hoped to reduce the risk of nuclear catastrophe by eliminating as many sources of human error as possible. So Nuclear Electric installed a software system of amazing complexity, consisting of 300–400 microprocessors controlled by program modules that contained more than 100,000 lines of code. The creation of such software is intuitive and not governed by any scientific laws. As Maner points out,

> this lack of governing law is unique among all the machines that we commonly use, and this deficiency creates unique obligations. Specifically, it places special responsibilities on software engineers for the thorough testing and validation of program behaviour. There is a moral imperative to discover better testing methodologies and better mechanisms for proving programs correct.

The task of testing 100,000 lines of codes is enormous and often such tasks are marginalised by the software engineers who, according to Maner, "test a few boundary values and, for all the others, they use values believed to be representative of various equivalence sets defined on the domain." Is this an ethically defensible approach particularly in safety critical systems?

A second example Maner uses concerns the speed of processing. On Thursday, 11 September 1986, the Dow Jones industrial average dropped 86.61 points, to 1792.89, on a record volume of 237.6 million shares. On the following day, the Dow fell 34.17 additional points on a volume of 240.5 million shares. It is believed that the drop was accelerated, though not initiated, by computer-assisted arbitrage. Arbitrage is traffic in stocks to take advantage of prices in other markets based on the spread: a short-term difference between the price of stock futures, which are contracts to buy stocks at a set time and price, and that of the underlying stocks. Computer systems constantly monitor the spread and identify when it is large enough to make a profit that more than covers the cost of the transaction between stocks and stock futures. Since the spread is small very large volumes are used in order to realise a worthwhile profit. The result is a lot of trading in a little time, which can markedly alter the price of a stock. After a while, regular investors begin to notice that arbitrage is bringing down the value of all stocks, so they begin to sell too. Hence the selling escalates and prices fall which is exacerbated by the computer's ability to process very quickly this enormous volume of dealing. The result is a stock market crash as was the case in 1986.

As Maner points out,

> The question is, could these destabilising effects occur in a world without computers? Arbitrage, after all, relies only on elementary mathematics. All the necessary calculations could be done on a scratch pad by any one of us. The problem is that, by the time we finished doing the necessary arithmetic for the stocks in our investment portfolio, the price of futures and the price of stocks would have changed. The opportunity that had existed would be gone.

But did the computer systems designers consider this? Did they understand the responsibilities they had in ensuring that their systems did not inadvertently destabilise the financial markets of the world. The answer is probably not.

Problems associated with the uniqueness of IT abound. Finally, consider these three statements:

■ Hacking is wrong.
■ Counselling support for the suicidal is right.
■ Anonymity between the counsellor and the counselled is a right.

And here is the dilemma:

■ Is it right to employ hackers to develop an anonymous internet counselling service for the suicidal?

What do you think?

4.3 New Horizons [1996]*

This section was first published as: Rogerson, S. (1996) ETHIcol—New Horizons. IDPM Journal, Vol 6 No 4, p. 15. Copyright © Simon Rogerson.

It appears a common trait of modern society that we are more concerned about the products of endeavours rather than the manner in which these endeavours are undertaken. In industry and commerce there often appears a common belief that the end justifies the means. The search for the best return and striving for competitive advantage dominate the decisions and actions of industry. This is certainly true within the information systems area where the pressure to deliver on time and within budget is often intense. A demanding economic climate has resulted in a short term perspective regarding survival and prosperity with little thought given to the long term wellbeing of a wider community.

It is generally accepted that the complex information systems development process is best undertaken using a project team approach. Effective project management is a vital ingredient in achieving a successful outcome. The planning element of project management lays down the foundations on which the project ethos is built. Here the scope of consideration is established, albeit implicitly or explicitly, which in turn locates the horizon beyond which issues are deemed not to influence the project or be influenced by the project. How the project is conducted will depend heavily upon the perceived goal. In his book, *How to Run Successful Projects*, Frank O'Connell explains that visualisation of this goal should address many questions including:

- What will the goal of the project mean to all the people involved in the project when the project completes?
- What are the things the project will actually produce? Where will these things go? What will happen to them? Who will use them? How will they be affected by them?

These types of questions are important because through answering them an acceptable project ethos and scope of consideration should be achieved. The problem is that in practice these fundamental questions are often overlooked. It is more likely that a narrower perspective is adopted with only the obvious issues in close proximity to the project being considered. The holistic view promoted by such questioning requires greater vision, analysis, and reflection. The project manager is under pressure to deliver and so the tendency is to reduce the horizon and establish a close artificial boundary around the project.

The scope of consideration is an ethical hotspot where activities and decision making are likely to include a *relatively* high ethical dimension. Within computing there are numerous activities and decisions to be made and most of these will have an ethical dimension. It is impractical to consider each minute issue in great detail and still hope to achieve the overall goal. The focus must be on the ethical hotspots which are likely to influence the success of the particular information systems activity as well as promote ethical sensitivity in a broader context.

The scope of consideration is influenced by the identification and the involvement of stakeholders both of which are often poorly executed. For example, when investigating 16 organisational IS-related projects Farbey, Land and Targett discovered that the perception of what needed to be considered was disappointingly narrow, whether it concerned the possible scope and level of use of the system or the range of people who could or should have been involved. Indeed, with the exception of vendors, all stakeholders involved were internal to the organisations and in close proximity to the projects. The implications of such restricted stakeholder involvement on achieving a socially and ethically sensitive outcome are obvious.

A wide range of stakeholders is associated with any organisation. Stakeholders are individuals or groups who can affect, or be affected by, the actions undertaken within an organisation. There are different sets of stakeholder which include:

- owners
- employees—both managers and workers
- unions
- capital providers
- government—at local, national, and international levels
- customers
- suppliers
- community groups
- competitors

It is from this list that participants are drawn. Participation by owners and employees is obvious but other groups may well participate in particular situations. If a manufacturing company was wishing to improve its supplier to customer chain then it would make sense to involve representatives from both groups. Similarly, if an organisation wished to form a strategic alliance with a competitor in an attempt to increase market share through synergy then participation by a competitor might be legitimate. The drive for efficiency gains through IS/IT application by a large local employer might potentially mean a reduction in the work-force or employment from a different work-force group. In such circumstances the involvement of unions and relevant community groups is probably desirable.

The widespread use of and dependence upon information systems within organisations and society as a whole means that the potential well-being of many individuals is likely to be at risk unless an ethically sensitive horizon is established for the scope of consideration of each information systems project. This horizon is more likely to occur if the principles of due care, fairness and social cost are prevalent during this activity. In this way the project management process will embrace, at the onset, the views and concerns of all parties affected by the project. Concerns over, for example, deskilling of jobs, redundancy, the break-up of social groupings can be aired at the earliest opportunity and the project goals adjusted if necessary.

No doubt there are those who will argue that such involvement will lead to extended debate which will be both costly and time consuming and is impractical in the techno-economic world of industry and commerce. However, such views are misguided and short term. The involvement of a wider group and the moving of the horizon as far away from the project as is practical will lead to a wider sense of ownership, commitment and acceptance of information systems. In this way information systems projects are more likely to be ethically sensitive by design rather than ethically sensitive by accident, a goal that all professionals must surely want to achieve.

4.4 Women in IT [1997]*

This section was first published as: Rogerson, S. & Stack, J. (1997) ETHIcol—Women in IT. IDPM Journal, Vol 7 No 6, pp. 7–8. Copyright © Simon Rogerson & Janet Stack.

This issue of ETHIcol addresses the role and treatment of women in the IT sector. Guest author, Janet Stack from the University of Glasgow discusses the problems facing women wishing to pursue a career in IT and the implications for the industry in failing to recognise and utilise this group of people.

The IT sector has experienced a massive expansion during the last two decades resulting in IT employees forming an increasingly vital component of the labour force. Alongside this expansion however the industry is now witnessing a severe skills shortage, which means it is all the more imperative that this industry maximises its use of all its potential human resources. Issues such as the millennium bug and monetary union have further exacerbated the situation.

Although there has been an increase in the number of women entering IT occupations since the seventies, women are still under-represented there in all member states of the European Union. This is an industry that has been predominantly white, middle-class and male-dominated with men more likely to be seen as the designers, developers and managers of systems whereas women are seen as the users of these systems. In no member state of the European Union is the proportion of women in the IT workforce estimated to be above 30% and in most cases it is closer to 20%. While all European Union member states report an under-representation of women, Spain and France consider that women are increasing their proportion in these jobs and will continue to do so in the future, while Germany, Italy, Belgium and UK see a declining trend. Out of all member states of the European Union, the UK has been experiencing he worst decline in the proportion of women in the IT sector. In particular, in the 1980s, women accounted for a quarter of the workforce in the IT industry in the UK. This had fallen to 19% in 1993, and this during a period when women account for about 45% of the UK labour force. The poor representation of females in employment is partly due to their low percentage of around 12% participating in a degree in computing. However, due to the increasing diversity of IT occupations there is increasing opportunity to enter the industry from a wide range of backgrounds. During the late 1940s and 1950s programming was the major type of computing work deemed suitable for women. By the 1990s, IT also includes jobs in the areas of systems management, communications and networks, systems development, technical support, operations, user support, and help-desk operations. The diversification of specialities and skills in this industry offer prospects not only for computing qualifications but also for those with business-like qualifications or post-graduate IT degrees to enter IT occupations.

The growth experienced by the IT industry has led to an occupational expansion both in terms of speciality and hierarchy. This is reflected in analysis of nation-wide surveys where specialities such as user support, communications and networks have been the fastest growing. The creation of new job titles emphasises the emergence of hierarchies in all specialities. Lack of job standardisation makes analysis difficult, but trends become evident and illustrate clearly that although women are found across all specialities they reach 40% only in areas related to administration and support functions. In programming and the more technical occupations the proportion reaches 25% but mainly in junior positions. Although women are visible across all specialities it is also evident that their presence declines steeply in more senior positions, both managerial and technical. Women who have reached senior positions are also disadvantaged in terms of salary.

The national statistics raise questions which need to be answered. Are these findings caused solely by women taking career breaks or part-time work for family reasons? Are supervisory positions only associated with full time employment? Does the industry regard it as a feasible proposition to institute policies for retaining these women in a career path? Is there equality of opportunity?

National surveys on gender differences in attitudes to careers in IT show that a higher percentage of females than males rate interest and challenge in their jobs as very important. Responsibility and on-going training are also rated very important by more females than males. Salary and location were the only factors which were rated very differently by gender: more men rated salary while more women location as very important. These surveys show that women have similar or higher aspirations than men.

Case study research in four varied organisations confirmed these findings. In one organisation studied, a woman had left after maternity leave because there was no part-time policy. As she stated, "flexibility to work from home was an issue that was discussed but never put into practice." One of the organisations did operate a successful part-time policy. In the words of a manager:

> All we want to see at the end of the day is that the job is done. So, we encourage all our people, we try to keep good staff by inventing ways for them to work. . . . technology can help you do this.

The company used voice mail and email to be in contact with employees who were not on site.

Stereotypes such as "males have a more burning interest towards technology" and "women are better in customer-focus," "women don't complain as much about the boring tasks assigned to them" and "part-time work makes it difficult for supervisory responsibilities" hinder women's progression in this dynamic industry. Lack of part-time and flexible work schemes and reluctance to use technology innovatively, all combine to deter retention. As a result, lower occupational positions, less challenging and less responsible work assignments are factors that not only affect opportunities for professional growth and promotion but also distance women from the exercise of authority and from positions of power.

The four organisations involved all made a conscious effort to support recruitment of females, yet none was identified as producing a gender-neutral environment. One manager emphasised the degree of organisation and planning required in software development but also the repetitive nature of some of the tasks: "I certainly find that some of the tasks that we are having to address as engineers, females are better at addressing them." As a female participant in the study recognised: "Characteristics that are admired such as open communication and collaborative approaches are not necessarily promoted." Employers may indeed prefer women to enter the field but they also tend to reserve certain positions, such as managerial and senior technical ones, for men thus perpetuating sex segregation within the IT industry.

Though the expanding and developing IT industry creates opportunities for women to enter these occupations, it also hinders their career development. The speed of change makes it difficult to catch up with technological advancements for those who take a career break, notably women who take maternity leave, and for those who can only work on a part-time basis, again mainly women with family responsibilities. Current policies and practices in the IT industry indirectly constrain women's development in these occupations. Male dominating attitudes and perpetuating stereotypes seem to pre-determine the positions that should be held by women who have managed to enter such occupations. Lack of innovative work practices and a career development path also hinder promotion. It is clear that retaining women in the workforce is an area of concern in addition to their recruitment.

There is immense value in having a diverse workforce. IT organisations with a workforce diverse in sex, race and age are more progressive in terms of innovative ideas and skills than those without. A diverse workforce introduces ideas and experiences into the workplace that may never be considered in a more traditional white male dominated environment. According to the IMIS report: "the fact that many of those organisations which are most imaginative in their use of IT to meet commercial objectives and/or consumer needs have a higher than average proportion of women in post suggest that the others are seriously losing out." There is a wider implication of the exclusion of women from higher-paid managerial positions, that where decisions are made on corporate IT strategy, the role of women to make an influence remains minimal.

To conclude, the growth in IT should have opened up new possibilities for women to enter these occupations. However, its growth has been used to construct and maintain gender differences and to create male hierarchies. Until women are more visible in all ranks of the IT industry, their current predicament will act as future deterrent to further recruitment so perpetuating the status-quo. The IT industry must increase women's participation and influence in both the strategic and development processes, rather than simply increase the total number of female recruits. Only then could it be considered to be a truly representative profession that promotes equality of opportunity.

4.5 IS in the Workplace [1998]*

This section was first published as: Rogerson, S. & Ivins, J. (1998) ETHIcol—IS in the workplace. IDPM Journal, Vol 8 No 6, p. 10. Copyright © Simon Rogerson & John Ivins.

In this issue guest author, Jon Ivins raises concerns over the implications of information flow in organisations and whether we really understand how information systems will change the workplace.

Information systems practitioners hold a range of views, usually optimistic, on the impact of new technology upon society. For example, in a recent interview, the head of BT Research, Peter Cochrane, argues that the advances in communication networks "will help banish ignorance and help the world community to become more cohesive." He explains that, "People talk about an Information Society, what we'll actually create will be an information economy," and warns about the need to ensure that everybody has the requisite ICT skills so that they can play their part in the new economy.

We all probably want to share such positive views of the impact of information systems on society at large. However, when it comes to the commercial implementation of IS into the workplace the new system is frequently seen as a way of doing the same things faster rather than an opportunity to do things better. The advent of electronic commerce, as discussed in the last issue, is indicative of how organisations must reinvent themselves to cope with new situations rather than simply focus on being more efficient.

There is often little or no evaluation of existing work practices and procedures in terms of their possible future relevance to the organisation. As a somewhat trivial example networked information systems allow businesses to dispense with routine tasks like the photocopying and filing of memoranda. They can be transmitted to the recipients and stored electronically. However, in many organisations the reality is somewhat different. Despite substantial investment in computers and networks the paper mountain continues to grow. Clearly the system can do the work but, given the continued use of paper, its appropriateness is called into question.

An example drawn from personal experience highlights the lack of understanding that many senior managers exhibit about IS. At a large organisation there was need for new information about a particular situation to be fed rapidly to staff. It was suggested that a private and secure, newsgroup could be set up to provide the dissemination of new information as it arrived. The head of the organisation refused to do this as "we are not an Email culture." This comment came after the manager in question had spent a large sum of money on the provision of a data network.

One issue that is frequently overlooked in a networked IS is the management and control of data generated by the staff. How will the data be stored, backed-up, maintained and kept secure? What checks are there to ensure that the corporate database holds coherent data rather than a collection of data useful only to individuals? Who can access the data? Do information management procedures fully comply with the appropriate legislation?

As well as data management, the move towards distributed processing gives rise to other issues. Decision making can take place at lower levels of the company hierarchy—indeed, recent years have seen a shift towards flatter management structures as technology has allowed work practises to be changed.

A flatter management structure clearly has implications as to how the flow of information takes place. Both formal and informal channels of communication are altered. Because of this the response time by the organisation and individual workers, to a given situation will change. Almost as an article of faith IS practitioners will point to an increase in the speed of processing information and argue that, consequently, an organisation will be able to respond much more quickly to changing circumstances. However, increased speed can be dangerous as the stock markets have found out to their cost. Indeed, they have amended their software to slow the response times down to avoid panic selling.

Informal information flows also play a vital part in many companies. They allow a frank exchange of "off the record" views, the sounding out of new ideas and give people a sense of being connected to the organisation and other workers. It is easy for IS designers and the managers of companies to forget to consider the impact of changes to informal information flows. Indeed, it is hard to identify them, or to measure their importance, in the first place. A few design methodologies do attempt to capture informal aspects of systems but many do not.

Many of the previous points have an ethical dimension. In general, the ethics of implementing a new IS are given little regard during the design process. Similarly, the health and safety of those using computers is given scant attention. Workflows, scheduled breaks from the computer and ergonomics do not seem to feature prominently in the design stage. This is not surprising when one considers that it is unusual for undergraduates studying IS and related subjects to be taught in any detail about the ethical implications of IS or relevant health and safety legislation.

Education must play an important part in addressing these issues. It must be an inclusive process drawing from both the IS user and IS development communities. Such cross-fertilisation of disciplines and ideas will help to lead us to Peter Cochrane's vision of a digital Nirvana. However, there is a long way to go yet before IS are truly empowering for both information workers and society as a whole.

4.6 Closer to Y2K [1999]*

This section was first published as: Rogerson, S. (1999) ETHIcol—Closer to Y2K. IMIS Journal, Vol 9 No 2, pp. 25–26. Copyright © Simon Rogerson.

As the year 2000 draws closer and closer organisations are becoming more frenetic in their efforts to ensure their computer systems are Y2K compliant. For example, around 30% of the 1999 IT budget of the top 100 European companies is being spent on Y2K compliance. There will be system crashes and there may be catastrophic outcomes resulting from these crashes. This is arguably the first worldwide IT problem that strikes all sectors simultaneously. It presents a unique and unprecedented threat to our civilisation which has evolved into a computer-dependent society and an interdependent planet that is fragile and susceptible to a domino effect when a computer system crashes.

It is the size of the problem that poses the greatest threat not the complexity. There are 180 billion lines of program code to check and 1 billion embedded microchips to locate and check together resulting in an estimated 700,000 person years of work. Some claim that there is not enough time or people to complete this work but there has been progress. The latest annual survey of IT and operational directors by research organisation Benchmark Research reports that only 2% of organisations have done nothing, 27% have been given verbal assurances from suppliers, 40% have completed a systems compliance audit and 20% have replaced systems to ensure compliance. Some organisations have been addressing the problem for some time. For example, Sainsbury's Board gave their Y2K compliance project the highest priority back in 1995 and since then they have been working flat out to get their systems ready.

Clearly this is a major problem and one which puts society at risk. So who is to blame for this and should the IT industry accept some of the responsibility? There are many in the IT industry that argue no. They argue that because of limited memory and costly storage it made overwhelming economic sense at the time to use two digit dates. Furthermore, software was not expected to last that long and so programmers did what was required to get a product up and working as fast as possible. The IT professionals of the day were simply ignorant of the potential problem at the turn of the millennium. Thus, two-digit dates became another de facto standard within the IT industry. Indeed, as late as 1995 microchips with two-digit date procedures were still being shipped.

Is ignorance a defence in this case? What does it take for the IT industry to be culpable for its ignorance? The IT industry ought to have known that two-digit date procedures would not work correctly beyond century boundaries. Indeed, there were examples of this regarding people born in the previous century and "special measures" having to be incorporated to cater for these "exceptions." These warning signs should have alerted the IT industry to the millennium problem. It can be argued therefore that IT professionals are directly culpable for their lack of knowledge resulting in the Y2K problem and hence are indirectly culpable for the resulting outcomes. Professionals should provide a higher order of care to the public and the IT industry is no exception. The public trust the IT industry to build reliable systems and it is questionable as to whether this trust has been abused in the Y2K context.

To be held responsible there must exist a causal condition and a mental condition. The former includes actions that were merely one significant causal factor among a number of others, while the latter includes unintended harm if the harm is brought about through negligence, carelessness or recklessness. If these definitions are accepted then there is a clear case of responsibility to answer because IT professionals acting in a careless and negligent manner were a significant causal factor of the Y2K problem. This careless and negligent manner included the lack of an integrated systems design approach, the lack of programming discipline and the lack of rigorous software engineering practices.

What has been the IT industry's response since the Y2K entry events of the creation and use of date algorithm routines? There were some early isolated reported concerns of the problem; in 1969/70 those working on systems in the UK in preparation for decimalisation identified problems with date field validation, in the US in 1976 software engineers working on a Medicare system discovered potential fraud opportunities because of Y2K and in 1989, the American social security administration discovered that payment scheduling would not work beyond 1999 because of Y2K. Such incidents appear to have gone unheeded and generally there was a lack of willingness to tackle the problem at an early stage. In fact there seems to be a powerful dynamic of secrecy. Leaders do not wish to panic their citizens; employees do not wish to panic their bosses; corporations do not wish to panic their investors and lawyers do not want their clients to admit to anything. As the veil of secrecy thickens, the capacity for public discourse and shared participation in solution finding disappears. The IT industry appears to be party to such secrecy regarding Y2K.

Since the Y2K problem has been anticipated a vast compliance industry has sprung up with large financial returns being made. IT consultancy firms are showering their Y2K workers with sign-on bonuses, loyalty bonuses, project bonuses, and high salaries according to a recent survey. Obviously the provision of a quality service in this area is worthy of appropriate financial reward but has the IT industry in some way exploited the situation? For example, is it acceptable to offer Y2K software upgrades for sale rather than provide free patches for what are effectively faulty goods? Similarly, is it acceptable that software tools that were not compliant were sold as late as 1997? Or is it acceptable that compression software that is suspect is shipped to customers? Organisations are concerned about their financial liability. It has been reported in the US that trade organisations are supporting a proposed new bill to limit liability of companies related to Y2K compliance. A recent news item reported that UK outsourcers were using termination clauses to avoid responsibility for Y2K crashes. Most contracts only have six or even three month notice periods so the impact on clients could be significant.

There are many serious questions to answer regarding the IT industry's role in creating the Y2K problem and its attitude in recent times to resolving this problem. A review of any of the codes of professional conduct in the industry suggests there might have been contraventions both in terms of attitude and practice. If so then the IT industry must have done something wrong, therefore it must be to blame, albeit in part, and therefore it can be held responsible. The industry should step forward, accept such responsibility and ensure that the risks of such worldwide problems occurring again are minimised.

4.7 Use and Abuse [1999]*

This section was first published as: Rogerson, S. (1999) ETHIcol—Use and Abuse. IMIS Journal, *Vol 9 No 3, pp. 26–27. Copyright © Simon Rogerson.*

There continues to be reported incidents of so-called computer abuse. Recent reports of a computer virus disabling many installations around the world, growing opposition from civil liberty organisations against workplace monitoring and the poorly designed electronic commerce system that allowed a child access to go on a mammoth shopping spree are examples of this growing problem. About a year ago the Audit Commission published their periodic report on computer abuse, titled "Ghost in the Machine." They reported a 9% increase to 45% over three years of organisations experiencing incidents of computer abuse. With organisations and society becoming increasingly dependent of information and communication technology (ICT) the upward trend of computer abuse incidents seems set to continue.

In contrast there are calls from governments to realise the potential of this technology for us all. In its introduction to the 1999 research and development work programme, the European Commission gives a clear indication that technology should be developed and applied in a wider social context. It states,

> The main focus . . . is on enhancing the user friendliness of the Information Society: improving the accessibility, relevance and quality of public services especially for the disabled and elderly; empowering citizens as employees, entrepreneurs and customers; facilitating creativity and access to learning; helping to develop a multi-lingual and multi-cultural Information Society; ensuring universally available access and the intuitiveness of the next generation of interfaces; and encouraging for all.

It explained that a wider focus must be adopted "in order to pay due consideration to the needs and expectations of typical users . . . in particular the usability and acceptability of new services, including security and privacy of information and the socio-economic and ethics aspects."

For such laudable goals to be realised there has to be greater emphasis placed on the ethical and social considerations relating to ICT practice. Computer professionals must address such matters urgently for they are the custodians of this powerful technology and as such are in a position of power and influence. They must recognise this privileged position and do whatever they can to reduce incidents of computer abuse. They must understand and accept their obligations as suppliers of systems, as users of systems and above all as members of the wider community. Clearly there are many issues to address some much more important than others. The sensible thing to do is to focus on those issues where there is a relatively high risk of ethical mistakes and where such mistakes have significant consequences for society, its organisations, or its citizens.

An ACM working group has been considering the practical management of computer ethics (albeit related to the university sector) for some time. Perhaps there is something to drawn from this for the ICT manager who must tackle these complex issues. The ACM working group proposed a set of rules of managing computer ethics (these can be found in a report available on the World Wide Web at John W Smith's website: www.people.virginia.edu/~jws3g/Publications/

PracticalManagementOfComputerEthics.htm). Those which are more applicable to the ICT manager include:

- make time to consider the ethical issues
- help your users understand the culture and hence the ethical stance
- when you do act regarding an ethical issue do so assertively
- do not let your manager get caught by surprise over an ethical issue
- remember that you have legal responsibilities which might affect the way in which you deal with an ethical issue
- do not give up your right to see what you need to see
- be consistent and fair
- do not discuss specific issues in public

Such statements might appear common sense and possibly vague, but they do put a different emphasis on the management culture within ICT. So often the focus has only been on the economic and technological feasibility of a proposal. These are obviously very important but so is ethical and social acceptance. It is not that ethics has been put aside rather that in the past it had not been recognised as a significant issue. This cannot be allowed to continue. As John Smith puts it, "Ethical problems and their potential consequences can be too serious to leave to chance. If you are responsible in any way for computing you cannot ignore these problems or leave them to others to deal with." ICT managers and their staff should embrace this mind shift and be proactive in self-regulation. If not then it is only a matter of time before governments impose stringent controls on the profession through, for example, legislation that requires computer professionals to hold a licence in order to practise.

4.8 E-mail Etiquette [2000]*

This section was first published as: Rogerson, S. (2000) ETHIcol—E-mail Etiquette. IMIS Journal, Vol 10 No 1, pp. 31–32. Copyright © Simon Rogerson.

There continues to be a spectacular global growth of electronic mail (e-mail) both in terms of messages communicated and people using such facilities. Indeed, there are now many people who cannot undertake their jobs unless they have e-mail facilities. Similarly, there are many people who depend on e-mail in their social lives. This is the latest ubiquitous communication tool and will soon be accepted as an everyday facility in the same way as postal and telephone services.

We do need to be concerned about how and by whom e-mail is used. For example, in December the New York Times sacked 20 employees at a Virginia payroll processing centre for violating corporate policy by sending "inappropriate and offensive" e-mail and the US Navy reported that it had disciplined over 500 employees at a Pennsylvania supply depot for sending sexually explicit e-mail. Such e-mail is clearly unacceptable and those involved should be dealt with accordingly but how was it discovered? Maura Kelly reported that some companies, including the New York Times, only check employee e-mail when they have been apprised of a violation of corporate policy, but others routinely monitor computer activities to identify employees who are slacking on the job or whose adult surfing habits or offensive e-mail messages could potentially expose a company to sexual harassment suits. Ethical and social problems associated with e-mail misuse will become more serious as e-mail expands. Such problems must be tackled.

Several have put forward suggestions regarding how to address e-mail ethics and promote good practice. In her book Netiquette, Virginia Shea suggests some rules which might encourage a socially responsible e-mail culture. These rules are summarised as:

Rule 1: Remember the Human
Rule 2: Adhere to the same standards of behaviour on-line that you follow in real life
Rule 3: Know where you are in cyberspace
Rule 4: Respect other people's time and bandwidth
Rule 5: Make yourself look good on-line
Rule 6: Share expert knowledge
Rule 7: Help keep flame wars under control
Rule 8: Respect other people's privacy
Rule 9: Don't abuse your power
Rule 10: Be forgiving of other people's mistakes

In New Zealand, the University of Massey has developed a code of ethics for the use of electronic mail. The code might form the basis of an approach for many organisations world-wide. Clearly it would need to be adapted to local needs and perceptions. The code's main components are:

■ forged e-mail. No electronic mail may be sent so as to appear to originate from another person, with the intention of thereby deceiving the recipient or recipients.
■ menacing e-mail. No electronic mail may be sent that is abusive or threatens the safety of a person or persons.
■ harassing e-mail. No electronic mail may be sent such that a person or persons thereby suffers sexual, ethnic, religious or other minority harassment or in contravention of human rights. The charge of harassment may be based on the content of the electronic mail sent or its volume or both.
■ privacy of e-mail. No person may access or attempt to access electronic mail sent to another user, without the permission of that user, except when necessary as part of that person's duties in respect of the operation of the electronic mail system.

(See www.massey.ac.nz/~wwits/emailpol.htm for further details.)

In many respects e-mail is a virtual frontier land. You need to be mindful of its pitfalls and traps. Never assume your e-mail messages are private nor that they can be read by only yourself or the recipient. It is very easy to forward an e-mail message to thousands of people. Unless you have complete trust that the recipient of your mail will keep it confidential, assume complete exposure to the rest of the world. It might be wise never send something that you would mind seeing it reported in the news media.

Everyone reading this article is likely to use e-mail. So based on your experience consider this interesting scenario written by Alpeda.

You are a computing officer in a reasonably large company with supervisor privileges on the computer. Over a weekend a crisis arises which demands attention and your superior phones in to ask you to get some information for her. She tells you that the vital sales report relating to a key contract that she needs is attached to an e-mail message sent between two of the directors on the previous Friday. She asks you to look at the message and forward its contents to her at home.

Would you do exactly as she asks, or do as she asks but with reservations, or refuse to carry this out and why?

4.9 Doing One's Civic Duty [2001]*

*This section was first published as: Rogerson, S. (2001) ETHIcol—Doing one's civic duty. IMIS Journal, Vol 11 No 2, pp. 25–26. Copyright © Simon Rogerson.

The new IMIS Code of Ethics was presented in the last edition of the ETHIcol (see Section 5.6). On reading this code it becomes clear that the computing professional has a wide range of demanding obligations to fulfil in the course of their working lives. It is not simply enough to comply with your contract of employment or commission. Being a professional is much more than this. One of the most challenging aspects is your civic duty. The code's first principle demands professionals to "uphold the health, safety and welfare of wider society, future generations and the environment." The demand is a proactive one. We must strive to ensure our work "will not be a cause of avoidable harm to any section of the wider community, present or future, or to the environment." We must "contribute to public debate regarding policy formulation in areas where this is in the wider interest." We must "oppose false claims made by others regarding the capabilities, potential or safety of any aspect of Information Systems and Information or Communication Technology."

It follows that professionals should do whatever they can to remove the digital divide. The statistical consensus is that only about 7% of the world's population is on-line. In global terms the dichotomy of "haves" and "have nots" is vast. With the potential benefit to society and its citizens this virtual divide should be a cause for great concern of all professionals.

Consider some of the facts. According to the US Internet Council's State of the Internet Report for the year 2000, there is considerable disparity within the continent of Africa in terms of internet accessibility. Africa has more than 10% of the world's population but less than 1% of the world's internet users. A million South Africans have access to the Web, but practically nobody does in the Congo.

Looking at South Africa in more detail, a 1999 survey of 2000 black South African men and 4,000 black South African women by Webcheck revealed dismal internet access rates among black Africans. The survey concluded that black African users of the internet have an access rate of about 1%. Out of those surveyed only two men and five women had Web access at home and only 24 men and 24 women had Web access at work.

Turning to India, recent analysis by Nua Internet Surveys and eMarketer reveals similar disparities. Indian e-commerce revenues are set to rise from $75 million at the end of 2000 to $254 million by the end of 2001, as cyber cafés enable increasing numbers of Indians to go on-line. The number of active adult internet users in India was 270,000 at the end of 1999 and had reached 1.5 million by the end of 2000, but this is only 0.2% of India's adult population. eMarketer attributes the low internet penetration in India to extreme and widespread poverty, compounded by a poor telecommunications infrastructure and limited PC ownership.

According to the report "Achieving Universal Access" by Booz-Allen and Hamilton, in the UK 14% of those unemployed and 35% of those in full time employment have internet access. This compares unfavourably with, for example, Australia where 39% of the unemployed and 57% of full time employed are connected. Looking at UK socio-economic groups; DEs represent 29% of the population but only 7% of the connected population whilst ABs represent 21% of the population yet comprise 48% of the connected population.

The report suggests that the divide will get worse. New users are currently joining the UK's on-line population at the rate of 4 million (or 8% of the population) each year. The report suggests that the UK market for internet access will sustain at least this growth rate over the coming years. At this rate, penetration in the UK should pass 60% "naturally" by 2003. However, far from evening out the emerging inequalities, the wave of growth is likely to exacerbate them in relative terms, leaving an unconnected or excluded group of over 20 million citizens.

These alarming facts demand action by individuals and organisations. Such action is in line with corporate social responsibility, a notion that is high on the general public's agenda. A 1999 landmark study, "The Millennium Poll," conducted by Environics International in co-operation with The Prince of Wales Business Leaders Forum, surveyed 25000 citizens in 23 countries regarding corporate social responsibility. The survey found that; 90% of people want companies to focus on more than profitability, 60% of people form an impression of a company based on its social responsibility which is defined as regard for people, communities, and the environment, and 40% of people talked negatively about companies that they perceived as not being socially responsible. The computing profession wants to be viewed more favourably by the public. It must understand that the public values social responsibility. Therefore, adherence to the First Principle of the code is paramount.

There are those who would argue that such regard is not in the interests of business but this is untrue. Such regard is not in conflict with your contractual responsibilities indeed a more socially responsible stance is good business. A 1997 DePaul University study found that companies with a defined corporate commitment to ethical principles do better financially than companies that do not. A recent longitudinal Harvard University study found that "stakeholder-balanced" companies showed four times the growth rate and eight times the employment growth when compared to companies that are shareholder-only focused. This holds true for all sectors including computing.

Organisations and individual professionals have a duty to act. If every computing professional was to contribute freely just 1% of his or her time to helping those on the wrong side of the digital divide just think what a difference it would make to social and economic justice and the restoration of reciprocity.

Note: The socio-economic classification produced by the UK Office for National Statistics is as follows AB: Higher & intermediate managerial, administrative, professional occupations; C1: Supervisory, clerical & junior managerial, administrative, professional occupations; C2: Skilled manual occupations; DE: Semi-skilled & unskilled manual occupations, Unemployed and lowest grade occupations.

4.10 Integrity and Knowledge [2002]*

This section was first published as: Rogerson, S. (2002) ETHIcol—Integrity and Knowledge. IMIS Journal, Vol 12 No 4, pp. 27–28. Copyright © Simon Rogerson.

We need reassurance that our electronic communications are safe from interference and delivered on time to the right people because ICT is not without its problems or its disasters. There continues to be increasing incidents of ICT abuse. For example, the Audit Commission has reported the incidence of ICT abuse within the UK every three years since 1983. The last two reports, Ghost in the Machine (1998) and yourbusiness@risk (2001) report an increase over six years from 34% to 67% of organisations experiencing incidents of ICT abuse. Such surveys suggest that unless positive action is taken the expansion of ICT usage will be matched by an increased level of ICT abuse. The consequences are manyfold; business disruption, damaged reputation, financial loss and a loss of confidence in the use of ICT in service delivery.

Clearly computer security must be addressed. If it is to be effective, its objectives must be related to business and be the responsibility of management. Senior management must be seen to support and be committed to the process. All managers and employees must be educated in and convinced of the need for security and be made fully aware of security policies and standards. There is a dual responsibility regarding ICT abuse. Organisations should minimise the opportunity for ICT abuse whilst individuals have a responsibility to resist such temptations.

Business operations that span the world and comprise multicultural workforces are now commonplace. Organisations frequently use outsourcing in the search for efficiency gains, so important in an increasingly competitive business world. This complex diversity in the workforce both in terms of backgrounds and relationships has added a new twist to the computer security issue. For example, where will the loyalty of the outsourcer lie? Will it be with the client, with the technology, with the outsourcing broker or with another third party? This new twist is particularly so given the increasing dependence of businesses on ICT both in internal operational activities and in the manner in which they interact with customers and suppliers.

Even with the most sophisticated computer security measures successful breaches still occur. It is people's attitude and behaviour that can make that extra difference. In the context of increasing globalisation, the moral attitudes and conduct of decision makers, providers, users and consumers of ICT are even more important. Achieving the moral high ground becomes increasingly more difficult in a global context.

People instigate security measures. All those involved in information systems are obliged to be on the lookout for computer misuse. We cannot abdicate this collective responsibility and assume that the computer security professionals will pick up this responsibility.

Computer security is as strong as the weakest link. If a neutral or amoral perspective is adopted then it will be the people who are the weakest link. The best technological security measures will be limited in such cases. Even worse is if an immoral perspective is adopted. Systems security is non-existent in such cases. This points to the need for a social contract between computer professionals and the society they serve. On the one hand society grants the right to practise as a professional, provides access to needed education, and passes necessary laws. In return the computer professional agrees to practise in a manner that benefits society. This includes ensuring as far as possible information systems are secure.

Consider the example of hacking which is a major security issue. The ethical question is whether or not hacking should be allowed or condoned. In answering this question Mikko Siponen suggests that we should place ourselves under a Rawlsian veil of ignorance where we are ignorant of our status, age, gender, and so on. The purpose of this veil is to foster impartiality. Behind this veil ask the question whether we accept that hacking is permissible so that anyone and everyone could break into our systems at any time. Clearly most of us under this veil would not accept hacking and would be vigilant in security enhancement and enforcement. This is because we would be concerned as to whether we are the victims of hacking.

Peter Neumann identifies three gaps that may permit computer misuse. There is a technological gap between what a computer system is actually capable of enforcing and what it is expected to enforce. There is a sociotechnical gap between computer-related policies and social policies. Finally, there is the social gap between social policies and actual human behaviour. Neumann argues that the technological gap can be narrowed through proper development, administration and use of computer systems and networks. The sociotechnical gap can be narrow by creating well-defined and socially enforceable social policies. Finally, the social gap can be narrowed by the actions for the other two gaps with additional support from education.

This education process must be twofold. It must include the education of computer professionals in understanding the impact of their actions on us all. This should lead to a realisation of the importance of such things as computer security. The second element of education must be associated with the public. Increasing public awareness of the use and impact of information systems would give the public confidence to question poor system implementation and demand improvement or withdrawal. The overall aim would be to educate people in the social cost and benefits of information systems. It would involve understanding the different levels of security and the need for different levels of security.

A culture of public confidence, trust of individuals and individual responsibility has to be developed, within which secure information communication can take place. There should be a commitment to training, strong support for good practice and an open management style to reinforce ethical behaviour within the organisation. An appropriate culture should be nurtured that is based upon a code of good practice which is accessible to all staff, visibly supported by senior management and has robust implementation procedures; an objective of enjoying a good reputation in society; and a strong tradition of dealing swiftly and firmly with transgressions. This is because, in the words of Samuel Johnson, "Integrity without knowledge is weak and useless, and knowledge without integrity is dangerous and dreadful."

4.11 Make Your Actions Count [2004]*

This section was first published as: Rogerson, S. (2004) ETHIcol—Make your actions count. IMIS Journal, Vol 14 No 6, pp. 29–30. Copyright © Simon Rogerson.

Information and communication technology is really about people not technology. It is a social construct. How people behave in developing and using the technology makes the difference in realising a society that is good for us all.

Two recent examples illustrate questionable actions by those who should know better!

The Financial Times on Wednesday, 6 October 2004 ran a story about new forms of mobile phone technology. It described a software product developed by Simedia called SounderCover which is designed to deceive when using mobile phones. The website for SounderCover urges you to "hide behind sound, make it your alibi:)." It continues, "Did you wake up late for work and you want your boss to think you're caught in traffic? Select the Traffic Jam background and give him a call from your bedroom:). He will hear your voice on top of this." (see www.simeda.com/soundercover.html).

The FT article quoted the founder of Simedia, Liviu Tofan as saying, "They [the nine background noises] sound very genuine and they give you the potential to pretend you're in a different place. We also give you a function which plays a telephone ring after 15 or 30 seconds, so you can say you need to get another call. Certainly people use it [SounderCover] to give plausibility to their excuses—both for work and in relationships." Sales for the product have soared.

It is hardly surprising that such products have become popular in a society where traditional values are put aside in the quest for celebrity and the obsession with reality entertainment.

This action by Simedia is clearly antisocial. Sometimes software product designs inadvertently provide opportunities for users to exploit its functionality in pursuit of unacceptable actions and goals. But this is different. Simedia are marketing this software on a ticket of deception. They want us to deceive. They want us to lie. To advocate such things challenges some of our core values such as truth and honesty. Such action is unacceptable and unprofessional.

The second case concerns the UK Passport Service (UKPS) which is currently running a six-month trial to test the recording and verification of facial recognition, iris and fingerprint biometrics. The claim is that the results will help inform the Government's plans to introduce biometrics to support improved identity authentication and help prevent identity fraud. These issues have been covered in previous ETHIcol columns.

The author of this column had the opportunity to participate as a volunteer in the trial. In doing so name, address and age had to be divulged but assurances were given that this personal data together with the biometric data would be destroyed after the trial. The trial consisted of two stages, the first where biometrics were recorded and a mock identity card produced and the second where this card was tested as an means of electronic identification. By chance the next volunteer was a person in a wheelchair. She was ushered into the first stage booth. The official assumed that given she was wheelchair bound she required the personal data on the computer screen to be read out to her so she could verify its correctness.

Given the proximity of the booths for the two stages, the author of this column, whist undertaking stage two, was able to hear all Margaret's details being read out. He challenged the official that this trial was in fact illegal as personal data was being divulged to a third party without consent through the lack of proper procedures. What was concerning was the response to this challenge. In the first instance the official failed to understand the breach of privacy that had taken place regarding the data subject Margaret. On further explanation his response was to ask the author to write a comment on the feedback form that volunteers were asked to complete. He seemed totally disinterested in this breach of privacy. Such actions seem to contravene the published privacy statement of UKPS (see www.ukpa.gov.uk/privacy.asp). It states, "We will take every precaution to protect your data" and "Your data is protected from unauthorised or accidental disclosure." Both were breached in this instance.

Given the sensitive nature of smart identity cards based on biometrics the unprofessional conduct of these officials is disturbing. If they fail to recognise and uphold the rights of citizens in trials what will be the attitude when this technology is fully implemented by UKPS? Use of technology, particularly in new situations, requires people to trust it. This unprofessional conduct does little to instil a sense of trustworthiness of this system by the general public.

These two cases highlight questionable behaviour by both developers and authorised users of technology. If technology is about people then those involved must be seen to be beyond reproach. Such people must exhibit virtuous behaviour. Indeed, DiPiazza and Eccles in *Building Public Trust* suggest that to secure public trust organisations must exhibit a spirit of transparency, sustain a culture of accountability and employ people of integrity. Would the organisations in the two examples pass these three tests?

As can be seen from the UKPS example organisations will have policies to cover good practice. Many will acknowledge they subscribe to a professional code of conduct such as the IMIS Code of Ethics. But that is not enough. In his article "Business Ethics for IT Managers-What You Can Do," Clinton Wilder argues that organisations must, "Be proactive. It's not enough to have a great ethics policy that sits on a shelf with the corporate mission statement. Institute regular ethics training and awareness programs. Move ethics away from "rules to be followed" to becoming a way of doing business."

The message is simple for IS professionals in all contexts—Actions speak louder than words—act ethically when developing and implementing ICT.

4.12 Risky Business [2005]*

This section was first published as: Rogerson, S. & Gotterbarn, D. (2005) ETHIcol—Risky Business. IMIS Journal, Vol 15 No 3, pp. 33–34. Copyright © Simon Rogerson & Donald Gotterbarn.

On 10 May 2005 Helen Beckett reported in Computer Weekly that, "Most IT directors do not know the full business risks associated with delivering IT services, according to a survey of

178 UK companies. The research found high levels of ignorance and low levels of confidence among IT managers in a critical part of their job. Some 60% said they were unsure how accidents or failures resulting from changes such as IT upgrades or office moves would affect the business." The findings of this survey are indicative of a widespread problem in every country where IT is used by organisations.

Here are five real examples of risky systems.

- A database system used to collect nationwide property data to identify trends in house prices and sales. There was a flaw in underestimating the significance of data integrity for profiling housing markets.
- A Maori genealogy database system was used in asset/wealth redistribution. There was a lack of recognition that this was socially sensitive data related to living people.
- An administration system was used in the healthcare management of senior citizens. The design was excessively technical and was unfit for purpose at best and totally unusable at worst.
- Electronic voting was proposed for use in national political elections. The proposal had over 100 problems associated with acceptability, usability and security.
- A data warehouse system was developed by a major software company. The system contained procedural errors which would result in major financial losses for the company.

Such risks are all ignored by traditional quantitative risk assessment used for IT projects and the purely quantitative approach to risk fails to alert developers to significant negative project impacts. Such risks cannot be turned into the quantitative measures related to time or money so often expected by organisational management in their desire to understand the risks associated with the latest systems development project. Quite simply systems development is a risky business that is out of control. Part of the problem is a narrow approach to risk assessment.

Research and practical project experience have shown that limiting risk assessment to purely quantitative risk analysis does not address a broad range of project risks including social, professional and ethical negative project impacts. A new type of qualitative risk analysis, based on engineering environmental impact statements has been developed to complement purely quantitative approaches to software risk. The Software Development Impact Statement process (SoDIS) extends the concept of software risk in three ways: it moves beyond the limited approach of schedule, budget, and function, it adds qualitative elements, and it recognises project stakeholders beyond those considered in typical risk analysis.

As the types of risks increase the range of stakeholders that need to be considered also expands. Using this expanded risk analysis reduced or eliminated the impacts of many previously undetected risks of software development. The successes of the SoDIS process, in the examples cited previously, provide strong evidence that a significant side-effect of narrowing project objectives and ignoring their social ethical and professional impacts is a root cause of the failure to understand risk properly (as reported by Helen Beckett) resulting in IT project failures.

The primary focus of developers is on the project development vision defined in terms of budget and schedule overruns and not on satisfying the customer by meeting technical requirements. The common intra-project risks are evaluated and managed using quantifiable values. The result is that the IT industry is still littered with project failures which are due in part to an institutionalised narrowing of the scope of a project's objectives and of the vision to development objectives. This explains why Beckett reports that, "Following change, 72% of IT managers were not completely confident that their business had a good understanding of the new technology assets installed."

Frequently the primary goals of project development—satisfy the customer, deliver the project within budget and deliver the project on schedule—are reversed. The focus of the risk analysis and mitigation narrows to those many issues which impact these goals negatively and risks that would derail the project's development. Often systems are evaluated in terms of the number of faults per 1000 lines of code rather than the side-effects these faults may have on system users or those affected by the system. These may be interesting numbers but they are totally misleading in their specificity.

Research done in the UK, New Zealand, and the US indicates that these inherent problems of existing risk analysis can be addressed in several ways including:

- expanding the list of generic risks
- maintaining focus on the broader project goals
- extending the list of considered project stakeholders

For a development project to succeed, risk resolution should consider:

- the delivered project type, consisting of sector, and application
  - the sector within which the software will be used
  - the type of application that is to be addressed
  - the surrounding circumstances of the application
- project impacts on all stakeholders
  - direct stakeholders—developers, customers, and others with a business interest indicating intra-project risks
  - indirect stakeholders—users, others whose life circumstances may be impacted by the product, and the social and natural environment indicating extra-project risks
- the different stakeholder expectations regarding how to judge a project as a success or a failure

Responsible risk analysis requires categorisation and description of the delivered project, and the associated direct and indirect stakeholders. The extension of risk analysis to a broader range of stakeholders is a necessary but not sufficient condition of adequate risk analysis. Limiting analysis to purely quantifiable risks would still miss many potentially negative impacts which cannot be easily quantified. There is a need to also focus on the broader impacts of the software.

The SoDIS process provides a mechanism for expanding the stakeholders considered and places particular focus on qualitative risks through the use of structured questions. The SoDIS process has been tested on software development in organisations with different location, size, function, scope, development methodology, and technology level; from small projects in consulting companies to projects as large as the UK's scheme for electronic voting. In one case, risks were identified which could have saved the company $250,000 USD.

The SoDIS process consists of four stages:

1. The identification of the project type together with immediate and extended stakeholders in a project.
2. The identification of the tasks in a particular phase of a software development project.
3. The association of every task with every stakeholder using structured questions to determine the possibility of specific project risks generated by that particular association.
4. Completing the analysis by stating the concern and the severity of the risk to the project and the stakeholder, and recording a possible risk mitigation or risk avoidance strategy.

The SoDIS audit process identifies significant ways in which the completion of individual tasks of an IT project may negatively affect any of the project stakeholders. It identifies additional project tasks and changes in existing project tasks that may be needed to prevent any anticipated problems. The resulting document, which complements a quantitative analysis, is a software development impact statement which presents all types of potential qualitative risks for all tasks and project stakeholders.

The SoDIS is the missing element in current risk analysis which primarily focuses on some of the quantitative intra-project relationships between selected tasks and selected stakeholders that constitute an IT project. A responsible professional risk analysis examines both the quantitative and qualitative associations between tasks and project's internal and extended stakeholders. To leave out either the quantitative or the qualitative analysis results in unidentified and, worse still, unaddressed risks and project failures. In efforts to continue improving the SoDIS risk assessment process and educate practitioners, the Software Development Research Foundation (www.sdre search.org) regularly works with companies to audit software projects.

4.13 All-Inclusive Opportunities [2005]*

This section was first published as: Rogerson, S. & McPherson, M. (2005) ETHIcol—All-inclusive Opportunities. IMIS Journal, Vol 15 No 4, pp. 33–34. Copyright © Simon Rogerson & Maggie McPherson.

The notion of inclusivity in an educational context is not a new one. As far back as 1792, Condorcet (as cited by Bown, 2000), stated that education "must be general and include all citizens" and that it should "insure that people at every stage of life have the facilities to preserve and extend their own knowledge." This is very much in accordance with current thinking in terms of widening participation and notions of lifelong learning. However, despite that fact that we have made great strides in the right direction, we are by no means there yet. That is probably why continuous professional development in IS/IT has not taken full advantage of these facilities particularly when the service provider and learner are geographically distant.

E-learning has been proposed as a means of reaching out to learners that previously were denied access to learning. Yet even a cursory glance at discussions on the advances afforded by technological innovations indicates that this is a hotly debated topic. According to Whitworth (2005), e-learning is becoming an increasingly integrated part of both the cultural and technological environment of the modern university, but this constitutes a significant alteration in these environments. Indeed, making use of technology within the learning environment represents a considerable change in terms of educational practice and culture, and the pace of change introduces a new level of complexity for both staff and students.

Up to now, some e-Learning developers and designers seem to have made the assumption that everyone has access to a connected computer, which in reality is far from the case. Fortunately, there is increasing evidence that mobile computing can alleviate some of the constraints of personal computers and limited bandwidth by moving directly into satellite services, wireless and mobile devices. For example, many Asian countries (spurred on by problems caused by SARS) have adopted this strategy and skipped some of the access problems related to e-learning and, as a result, have a significantly higher rate of usage of educational technologies than in North America and Europe. Trotter (2005) reports on a plan in progress to wire up to 300 universities in the Asia Pacific area by the end of next year. The intention is to allow students and lecturers to experience a more flexible learning and teaching environment by providing remote access to educational resources, including research libraries. Even countries that lag behind in bridging the digital divide are catching up in this regard. For example, India launched EDUSAT, its first satellite dedicated to education services, late last year.

Nevertheless, access is not the whole issue of inclusivity. Hoffman and Novak (1998) asserted that a combination of low income and education contributed to the lack of access of a number of groups as largely identified as "undigital." Thus, it should be noted that simple digitising of educational content will not necessarily address the problems faced by those who have neither the technical skills nor the ability to take part in opportunities afforded by e-learning. Many groups, such as the poor (especially in rural areas), those from ethnic minorities, and single parent households, are not thought to be in a position to take advantage of these wonderful new educational opportunities. This then compounds an existing socio-economic divide between the "haves" and the "have nots" of the world, and this is an issue that does not affect the "developing world" alone.

Even for those fortunate enough to be included in the digital society, unlimited access to information poses a very real problem. Over 30 years ago, Alvin Toffler (1970) signalled that we faced an overload of information and termed this "future shock syndrome." His theory was that the human brain could only absorb and process a finite amount of information, and if exceeded, thinking and reasoning becomes adversely affected, decision-making flawed and, in some cases, impossible. No wonder then that, when faced with information of which they are unable to make sense or fail to see its usefulness, many people today feel overwhelmed and even become ill.

Despite the take-up of educational technology within many colleges and universities, early adopters of e-Learning have often created courses based on the norms and practices of traditional boundary-limited campuses. To provide a sound basis for an educated society, it is necessary to turn information into knowledge and it is important to ensure that students of today learn how differentiate between what is valid and useful and what is irrelevant and unsubstantiated. This demands the ability to sift and manage information as never before. Thus, e-learners not only require the development of skills of discernment of content and quality, which is best achieved through active learning, but also need easily assimilated and suitable mechanisms for information management. In this way we will avoid the promotion of inappropriate bite-learning.

Since Virtual Learning Environments (VLEs) comprise the components in which learners and tutors participate in online interactions of various kinds, including online learning (Jisc, 2003), VLE developers and e-Learning course designer have a vital role to play in ensuring the finest design quality possible, which will encourage academic staff and students to make best use of these new opportunities. Thus, the challenge is to design VLEs with all the necessary resources built in for active learning, and still ensure that these are really accessible, usable and adaptable for tutors and learners that need them.

If these challenges can be met then the possibilities new professional development offerings abound for IS/IT professionals.

4.14 Women in the IT Industry Today [2005]*

This section was first published as: Rogerson, S. (2005) ETHIcol—Women in the IT Industry Today. IMIS Journal, Vol 15 No 6, pp. 29–31. Copyright © Simon Rogerson.

Two articles about women in IT appeared in the press recently. Writing in *Computing* on 8 September, Gillian Arnold explained that, "Recruiting and retaining experienced women in our industry is of paramount importance, and is fundamental to the future competitiveness of the UK's IT industry. In an industry that regards itself as leading-edge in the value it delivers to the

UK economy and society, companies can no longer avoid addressing their failure to create a working environment that attracts and retains senior, qualified women. The number of women working in the IT industry fell from 27 per cent in 1997 to 21 per cent in 2004. And even though the number of women employed in the IT industry has stabilised, the sector is still losing experienced, senior women from its ranks. Unfortunately, there are still organisations, or senior individuals, within the technology industry that believe talk of diversity is a meaningless diversion from the conduct and professionalism of everyday business life."

By contrast *Female IT managers topple men in pay league* appeared on the silicon.com website on 19 September in which Andy McCue reported that

> Female IT managers are for the first time earning more than their male counterparts and women are also climbing the corporate ladder faster, according to the annual salary survey from the Chartered Management Institute (CMI). Across all sectors the average female team leader, at 37 years old, is four years younger than her male counterpart, while female IT managers earn on average £45,869 per year—£779 more than the men do.

It was reported Paul Campfield, director of Remuneration Economics, said in a statement:

> It is encouraging to see that the number of female managers continues to increase but it is worrying that they are still more likely to resign. The implication is that female managers still face difficulties in the workplace and organisations should address these quickly because, unchallenged, these problems will demotivate and disrupt with the end result being poor performance and productivity levels.

So, what is the truth about women's prospects and influence within the IT industry? Several experts have given their opinion. Professor Alison Adam (UK) pointed out that,

> All the evidence so far suggests that women don't fare as well as men in the IT industry in financial terms—things like salary secrets are rife. . . . I doubt this [CMI survey] will mean that there has been a sudden change in the IT industry in gender equity terms-our research [at Salford University] so far suggests that women still have a pretty tough time in the IT industry.

Professor Wendy Hall (UK) of University of Southampton agrees, "it's almost certainly true that less than 20% of the IT workforce is female. And very few of these women get to the top of this very macho world. So those that do get to the top have to be very good." Similarly, Professor Teressa Rees (UK), Pro Vice Chancellor Student/Staff Issues at Cardiff University explained that, "There is a pay gap among [IT] graduates within three years of graduating, companies may well recruit women but are not so good at retaining or promoting them. These figures [from the CMI survey] should not be used to invoke complacency."

In the US the situation for women in IT is difficult. Caroline Wardle (US) explained that,

> it is not unusual for women and men to be hired at the same salary levels on entry to high-skilled IT jobs, but after a few year the men's salaries outstrip the women's. It is also the case that women's representation in the US IT workforce has been dropping steadily over the past decade and a half.

According to the US Department of Labor, Bureau of Labor Statistics Report 985 produced in May of 2005, women computer and information systems managers earn 79.4% of men's earnings. For computer scientists and systems engineers, women earn 82.6% of what men earn. So the gap has not closed in the US.

stated Professor Fran Grodzinsky (US).

Experts went on to discuss the reasons behind men/women IT employment differences. Eva Turner (UK), Organiser of Women into Computing 2005 pointed out that;

> [McCue] does not indicate anything about the power struggles and the conditions under which all workers in today's ICT industries are expected to work (high pressure, highly intensive, available 24 hours/day, no unions etc.). . . . There is a documented trend of women top managers choosing not to start a family or have children as the pressure of the top success does not allow them to make that choice.

This theme was picked up by Professor Fran Grodzinsky (US); "Our research supports the idea that women are leaving when they have to balance family and child rearing with the enormous time commitment that a managerial role demands. A change in the corporate culture that builds in flexibility might alleviate the drain of top women talent from the industry." If this change does not happen then perhaps the observation by Vanessa Hymas Deputy CEO of IMIS will remain true;

> I wonder whether the surveyed females can spot more easily that the battle cannot be won, so move on to an organisation more appreciative of their worth, or whether other pressures on their lives suggest to them that life is just too short to continue the struggle!

Concerning talent and ability of IT women professionals, Professor Wendy Hall (UK) said, "Increasingly every aspect of what the company does will rely on IT, and so the IT manager/director role is increasingly important in terms of managing up as well as down. These are all skills that bat to the strengths of women." However, there was a different view from Professor Vivian Lagesen from the Department of Interdisciplinary Studies of Culture at the Norwegian University of Science and Technology. She was dismissive of the idea that women are better than men at people skills explaining, "and as my own research has shown, there is a widespread notion or discourse about women being better at communicating, having better people skills, being more versatile, etc. that makes them better computer scientists and perhaps better managers." She continued,

> Since many women seem to be not that obsessed with programming, as some men are, they are more likely to seek other career opportunities and often also chooses the way out of the most technical areas of computing and toward other areas such as sales, marketing and as this article shows, also management. Also, my own research shows that women to a much larger extent than men are recruited to computer science studies because of career opportunities, whilst men are more often recruited because of interest in computers. I think these may be the most important points in this respect.

4.15 Framing Up to Ethics [2006]*

This section was first published as: Rogerson, S. (2006) ETHIcol—Framing up to Ethics. IMIS Journal, Vol 16 No 4, pp. 29–30. Copyright © Simon Rogerson.

Much is made of the convergence of technologies with each other as well as with other fields. Such convergence and the associated blurring of boundaries seem to reflect the maturing of the underpinning science and technology together with a growing acceptance that real world problems and needs can only be addressed effectively using a pan technological-scientific-sociological approach. A good example of this is the manner in which computers, telecommunications and media have converged and are challenging traditional broadcasting and reporting. The open online personal journals of Blogs, the self-correcting, participatory media of Wikis and the recording and offline listening of Podcasts have led to citizen journalism which satisfies the needs of society to access alternatives to the manicured news offerings of the information barons.

From an ethical perspective commentary and guidance regarding the converging technologies is offered under computer ethics, communication ethics, information ethics, bio-ethics, nano-ethics and so on. For practitioners this may well appear distant, bewildering and irrelevant, particularly those involved in ICT whose role is to provide workable solutions and services.

However, in the modern technology-based economy, the role of the ICT professional is critical. Such professionals have differing relationships with the organisations they serve depending on whether they are employees, contractors, consultants, temporary staff or volunteers. Nevertheless, regardless of the relationship all have a critical role to play.

A recent report from the Australian National Training Authority revealed that in the ICT industry "an important message [was] about the significance of people, and the value of knowledge, creativity, foresight, and wisdom." Such valuation supports a proactive approach to encouraging professionalism within the ICT industry. Indeed, given the debatable success of ICT professionals to deliver fit-for-purpose systems it would seem that there needs to be improved frameworks or scaffold to assist this disparate group of employees, contractors, consultants, temporary staff and volunteers to give of its best in support of both the employing organisations and society in general.

The increasing ICT dependency of organisations and society requires demonstrable trustworthiness in ICT professionals. The difficulty is that trustworthiness is intrinsic. In order to judge the trustworthiness of a person we therefore resort to extrinsic cues such as smart appearance, a firm handshake, a shared joke, social capital, association and declaration. The framework or scaffold mentioned earlier could be a powerful extrinsic cue and could help sustain ICT knowledge, creativity, foresight, and wisdom within organisations. It might comprise, for example, a suite of guidelines and policies, a training programme, an online discussion board and information library and a staff rewards scheme.

It is important to recognise that the conditions must be favourable for establishing the framework or scaffold. There must be managerial support at the highest level. Stringent administrative control must be replaced with flexible and tolerant coordination. Radical, innovative views and perspectives must not be curtailed by traditional, conservative views. The framework or scaffold must provide aspiration and expectation. Without aspiration, expectation can become regulatory and tedious. Without expectation, aspiration can become high sounding but empty. Overall, the framework or scaffold will cover values, behaviour, processes and outcomes. The constituent parts of the framework or scaffold must be created and maintained by the ICT community with the help of specialist expertise. In this way the framework or scaffold will exhibit integrity and be owned by the working community.

Four interrelated topics need to be addressed in the construction of the framework or scaffold. The first topic is ICT work itself. How is the work defined in terms of jobs and activities? What are the ethical principles to uphold in the doing ICT work? How is quality addressed? What are the cherished practices to adhere to? The second topic concerns the authority under which ICT operates. What is this authority and what are the stated values of this authority? The third topic relates to the approach adopted in the framework or scaffold construction. How will the approach be cohesive ensuring it remains current and relevant? The final topic concerns identifying those who will build the framework or scaffold. What combination of ICT specialists, other specialists and administrators are needed in the construction? What will their roles be and what will the relationships be between these groups?

If such a framework or scaffold is constructed appropriately and used effectively then an organisation will increase the chance of ICT professionals being regarded as trustworthy. In turn trustworthy ICT professionals increase the likelihood of realising beneficiaries rather than victims of the systems developed through their professional endeavours.

4.16 The Need for Ethical Leadership [2007]*

This section was first published as: Rogerson, S. (2007) ETHIcol—The need for Ethical Leadership. IMIS Journal, Vol 17 No 4, pp. 29–31. Copyright © Simon Rogerson.

The use of technology requires strong ethical leadership if society is to benefit from research advances. Unfortunately, this does not always happen.

On 11 July 2007 at the launch of the annual report for 2006/07, Richard Thomas, the Information Commissioner, said: "Over the last year we have seen far too many careless and inexcusable breaches of people's personal information. The roll call of banks, retailers, government departments, public bodies and other organisations which have admitted serious security lapses is frankly horrifying." Indeed, in February 2007, the ICO found Alliance & Leicester, Barclays Bank, Clydesdale Bank, Co-operative Bank, HBOS, HFC Bank, Nationwide Building Society, Natwest, Royal Bank of Scotland, Scarborough Building Society, The Post Office and United National Bank in breach of the Data Protection Act and ordered them to sign formal undertakings.

Thomas continued:

> How can laptops holding details of customer accounts be used away from the office without strong encryption? How can millions of store cards fall into the wrong hands? How can online recruitment allow applicants to see each others' forms? How can any bank chief executive face customers and shareholders and admit that loan rejections, health insurance applications, credit cards and bank statements can be found, unsecured in non-confidential waste bags?

The Information Commissioner added: "Business and public sector leaders must take their data protection obligations more seriously. The majority of organisations process personal information appropriately—but privacy must be given more priority in every UK boardroom. Organisations that fail to process personal information in line with the Principles of the Data Protection Act not only risk enforcement action by the ICO, they also risk losing the trust of their customers."

When a technological advance is applied to resolve a real world problem or to provide a new product or service it often raises unanticipated moral implications. The lack of adequate safeguards leading to breaches in privacy illustrates this. This clearly points to a need for improved leadership.

In her 2005 Coca Cola Lecture "Ethical Leadership for the 21st Century" at Morehouse College, Dr Shirley Ann Jackson, President of Rensselaer Polytechnic Institute suggested there were six principles of leadership regarding technological innovation.

1. Integrity This "means holding to the highest standards and setting an ethical example—on a continuous basis."

2. Vision "There are two kinds—vision which sets the tone and the direction, and which must be rooted in a clear-eyed view of the big picture, and vision to delineate the complex, intertwined ethical questions, amid their larger context and their implications for society at large."

3. Courage "To lead requires the courage to make, and to stand by, difficult choices."

4. Engagement This is "of others in coming to a principled stand on a difficult issue, and in getting others to join you in staying the course. In addition, engaging constituencies and enlisting participation enable the decision-making to profit from multiple insights and creative energies."

5. Language "Language helps set the tone and can include—or exclude. Language has the capacity to elevate the discussion, drawing the best from participants."

6. Action "One must be willing to act, not merely to believe. Action is needed to fashion practical approaches"

Such principles can be translated into practical organisational policy and initiatives. For example, Robert H Dunn speaking at the Ethical Leadership Forum 2000 in Hong Kong suggested that ethical companies involved in internet business should have at least the following characteristics.

■ "A strong statement of company values, an ethical framework applied to all business decisions, and an explicit acknowledgement that all aspects of business operations, including ethical values, apply in cyberspace."

■ "An ethics-based code of conduct that goes beyond legal compliance, and gives specific guidance on cyber ethics challenges."

■ "High visibility support for the ethics program on the part of senior managers."

■ "Effective, ongoing communications and training that address the special ethical challenges of operating in cyberspace."

■ "Helping resources to support employees in making good and timely decisions."

■ "Clear rewards and sanctions that hold people accountable."

■ "Some system of stakeholder engagement to ensure that you are adequately addressing and balancing the ethical concerns of all those impacted by your decisions."

In her lecture, Jackson concluded:

In today's wi-fi, hot-wired, hyper-linked, 24-7-365 on-line world, however, leadership challenges are no longer single, or simple. They intertwine and reach beyond a single community to touch communities across the entire planet. Ethical leadership in this environment ultimately is answerable to a global community. Leadership which is effective must take into account a vast array of needs, capacities, skills, backgrounds, perspectives, cultures, languages—and, yes, hopes. For it is hope which ultimately lifts the human spirit, and motivates people to take ethical stands and to act on them." The challenge is to address this positively so that organisations will, as Dunn points out: "explore ways they can harness new information technologies to promote ethical behaviour—not simply to build a firewall against the risks that technology creates.

4.17 Ethical Dilemmas [2009]*

*This section was first published as: Rogerson, S. (2009) ETHIcol—Ethical Dilemmas. IMIS Journal, Vol 19 No 2, pp. 28–29. Copyright © Simon Rogerson. NOTE: Some of the dilemmas in this article are updated and modified versions of some that appeared in the Multi-National Survey of Information Systems Ethical Perceptions conducted by The University of Nevada, US in 1996.

The professional world of ICT is littered with ethical dilemmas. Sometimes the ICT professional will recognise such dilemmas and strive to resolve them. Other times such dilemmas will be ignored and might be considered outside the employee's jurisdiction. Often the ethical dimension of ICT is simply not recognised. All ICT professionals need effective training on how to address these issues in a way that leads to acceptable professional practice and consequential acceptable products and services. Dilemma training should be part of the continuous professional development of the ICT professional regardless of job type or seniority.

Consider the following scenarios. How ethical or unethical do you consider the situations to be? What would you do in this situation? What would you do if you supervised the person(s) involved in each situation?

Scenario A—A computer programmer built small computer systems in order to sell them. This was not his main source of income. He worked for a moderately sized computer vendor. He would frequently go to his office at the weekend when no one was working and use his employer's computer facilities to develop his own systems. He did not hide the fact that he was going into the building; he had to sign a register at the security desk each time he entered.

Scenario B—An employee provides network and applications support for users throughout the organisation. One day, the employee was undertaking a major upgrade to all the laptops within the company. While working on the laptop of the Finance Director, the employee noticed that this laptop contained thousands of pornographic pictures.

Scenario C—An employee fills out a survey form on a computer game web page using her work's computer whilst on her lunch break. The survey asks for her email address, mailing address, and telephone number which he fills in. In the following weeks, she receives several advertisements in the mail as well as dozens of email messages about new computer games.

Scenario D—A trainee at a large finance company learned to use an expensive specialised spreadsheet program in her accounting work. The trainee would go to the finance company's computer suite, book out the CD-ROM holding the software, complete her training tasks, and return the software. Notices were posted in the suite indicating that copying software was forbidden. One day, she decided to copy the software anyway so she could work on her tasks at home on her own computer.

Scenario E—The project manager suspects a team member of using his company email account to send offensive messages to other employees of the company. She asks the company's network manager to give her copies of team member's email.

Scenario F—A mobile phone company employee saw an advertisement in a newspaper about a car for sale. The car sounded like a good buy to the employee. The advertisement listed the seller's mobile phone number, but not the seller's address. Being a system software engineer, the mobile phone company employee knew he could determine the seller's address by accessing the seller's mobile phone records. He did this and went to the seller's house to discuss buying the car. The seller was delighted and the sale went through.

Scenario G—An employee suspected and found a loophole in the organisation's computer security system that allowed him to access other employees' records. He told the system administrator about the loophole, but continued to access others' records until the problem was corrected two weeks later.

Scenario H—An engineer needed a program to perform a series of complicated calculations. She found a computer programmer capable of writing the program, but would only employ the programmer if he agreed to share any liability that may result from an error in the engineer's calculations. The programmer said he would be willing to assume any liability due to program malfunction, but was unwilling to share liability due to error in the engineer's calculations.

Scenario I—An ICT professional is employed by a large hospital to maintain and upgrade its patient database. Her aunt, who works for a medical insurance company, approaches her one day to help her get the medical histories and other personal particulars of patients of the hospital who had died of a certain illness, so that her company can formulate a new insurance scheme for such patients. She explains that this will offer a much better financial benefit than currently available for spouses and dependants.

So next time you are at the coffee machine talking with colleagues why not raise one of these scenarios with them and see how they react. Discussing ethical issue related to work in an informal setting is a good way to promote an ethical culture which raises awareness and helps everyone deliver acceptable products and services.

4.18 Coding Ethics into Technology [2017]*

This section was first published as: Rogerson, S. (2017) Coding ethics into technology. Hack & Craft News, *3 July, Additional reporting by Chris Middleton, http://hncnews.com/coding-ethics-technology. Copyright © Simon Rogerson.*

As a young Fortran programmer in the 1970s, I was once told to incorporate a covert time bomb into a design system that was to be rolled out to a subsidiary company. At the time, I saw nothing wrong in building these functions; the ethics of the decision didn't cross my mind. After all, I was a junior programmer and had been told to do it by the most senior member of staff in the department. You might say I was an uneducated technologist.

Today, I believe that professional practice is unprofessional without ethics, and yet it seems that little has changed in the industry. In January 2017, for example, car giant VW pleaded guilty to using a defeat device to cheat on mandatory emissions tests, as well as to lying and obstructing justice to further the scheme.

At the centre of this scandal was misinformation generated by onboard software. That system was developed and implemented by computer professionals, who must have been party to the illegal and unethical purpose behind it.

Nearly half a century passed between these two events, which suggests that the software industry has learnt little about the importance of ethics in system design. But as technology becomes more and more central to our lives, the ethical dimensions ought to become more central too.

For example, surveillance is rarely out of the news, and it is often said to have a moral purpose: to catch terrorists and abusers. Yet in the US, it has been revealed that the FBI is designing a system to gather, identify, and (via AI) contextually analyse images of tattoos—not only to identify individuals by them, but also to infer the meaning of any tattoo within the wider population. Such a system may flag innocent people as suspects—and potentially as people who share the same values and beliefs as other suspects.

Is that an ethical development? Should technology professionals help society to walk down that road? And does the very real scientific problem of confirmation bias come into play here:

designing systems that confirm pre-existing prejudices? (That would seem to be self-evident in any system that links tattoos with criminality.)

Such questions are important, because IT professionals' unethical decisions aren't always *deliberately* so: sometimes they arise from too little consideration of their own frames of reference.

Take this example: at the World Economic Forum 2017 in Davos, MIT Media Lab's Joici Ito revealed how a group of students had designed a facial recognition AI system that couldn't recognise an African American woman. This wasn't because they were consciously prejudiced, but because they hadn't considered that the development and testing environment was exclusively white and male—as is common in the industry. The system was released before anyone spotted the problem.

In this example, no one stepped outside of their own frame to consider the project from a different angle. "The world outside the box" can—and should—present an ethical perspective. Ito's anecdote demonstrates how a lack of diversity has both ethical dimensions and a real-world impact.

IT development can reinforce societal problems if it fails to consider them at the design stage. But is the industry trying to fix the problem?

In 2014, the BCS, the Chartered Institute for IT in the UK, ran a special edition of ITNOW focusing on ethics in ICT. In many ways, it was a litmus test of ethics progress by practitioners and academics working in tandem across the industry. It was a disappointing read: the lack of ethical consideration in systems design and implementation was evident, and the calls for action were neither new nor inspiring. There was virtually no evidence supplied and no pragmatic action demanded; the emphasis was all on top-down political rhetoric (see Section 7.15 for a detailed analysis of this edition).

The report illustrated that, at best, the industry has stood still.

Ethical practice should be paramount among computer professionals. But what does this actually *mean*? Practice has two distinct elements: process, and product.

- Process concerns the activities of computer practitioners and whether their conduct is virtuous.
- Product concerns the outcome of their professional endeavour and whether the systems are ethically viable.

Time bombs and defeat devices fail on both counts, while racist AIs and invasive surveillance systems fail against the second (arguably, in the case of the FBI example). But why should IT professionals care more about these problems, and what can they do about it?

First the question of why.

Every day, society becomes more and more reliant on information and communications technology. Our innovations seem limitless, as does their scope to seep into all aspects of people's lives. Application areas such as the internet of things (IoT), cloud computing, social media, artificial intelligence (AI), and big data analytics are commonplace—not just in enterprise contexts, but also in everyday consumer ones.

Some argue that, as a consequence, society becomes more and more vulnerable to catastrophe. Those fears are based on fact. For example, the ransomware attacks of May 2017—themselves designed by coders, of course—caused the closure of many hospital Accident & Emergency (A&E) units in the UK, and in June, caused worldwide disruption in banks, retailers, an airport, and energy suppliers, including a nuclear power plant.

In the commercial world, the drive for efficiency, productivity, effectiveness, and profit is seen as the priority by strategists and business leaders, and this affects what IT professionals are asked to do.

Such pressure sometimes results in real short-term gains, but it can also lead to unscrupulous, misguided, or reckless actions (as we have seen).

Sometimes those actions are really *inactions*. For example, it was revealed in the technology press that the main reason for the ransomware's "successful" takedown of hospital systems was due to the operating system being out of date and unsupported, because of cost cuts. The government had been repeatedly warned of the risk.

The tempering of efficiency/profit drives with greater ethical considerations of their outcomes is often neglected—until something happens and triggers a public outcry. As a society we seem to accept this, but computer professionals don't have to. Coder and technologists don't have to accept playing their own part in unethical behaviour, such as designing systems to fail or to mislead the public.

But what can the technology industry do about it? There exist several ethics tools which can be used in the design process of systems. Three of these are DIODE, FRRIICT, and SoDIS.

DIODE is a structured meta-methodology for the ethical assessment of new and emerging technologies. DIODE was designed by a team of academics, government experts, and commercial practitioners to help diverse organisations and individuals to conduct ethical assessments of new and emerging technologies.

The Framework for Responsible Research and Innovation in ICT (FRRIICT) is a tool that helps those involved in ICT R&D to carry out their work responsibly. It consists of a set of scaffolding questions that allow researchers, funders, and other stakeholders to consider a broad range of aspects of ICT research.

The Software Development Impact Statement (SoDIS) extends the concept of software risk in three ways. First, it moves beyond the standard limited approach of schedule, budget, and function; second, it adds qualitative elements; and third, it recognises project stakeholders beyond those considered in a typical risk analysis.

SoDIS is a proactive, feed-forward approach which enables the identification of risks in how ICT is developed (and within ICT itself). It is embedded into a decision support tool, which can be used from the start of system analysis and design.

Why such tools have not been incorporated more into system design is open to question. Perhaps it stems from a mismatch between the pressure to complete quickly and cheaply, and the obligation to complete properly. (We can all think of applications that are rushed to market, then fixed on the fly later. In such an environment, ethics take a back seat.)

Or perhaps it's a symptom of inappropriate education and training, where the focus is firmly on the technology at the expense of the context.

Or perhaps it's a symptom of "silo thinking" across practitioner and research communities, which prevents valuable exchange and synergy.

Or perhaps there is a general lack of awareness of the dangers associated with technology, especially among coders who prefer the binary world of computers to the messy world of human beings—an observation made by MIT's Ito at Davos.

Whatever the reasons, the situation must change. All those in the computing profession, including new entrants, need to have the ethical tools, skills, and confidence to identify, articulate, and resist the unethical aspects of system design and implementation.

More, they should be free to challenge the decisions of, and orders issued by, their seniors, where those actions are ethically questionable, without detrimental effect to themselves.

There should be "three E's" in technology development: effectiveness, efficiency, and ethics, and these should be applied right from the start of any programme.

The tools to do this exist, but it is the desire that seems to be missing. And that's why the charge of professional irresponsibility within the computer profession cannot and should not be ignored.

References

Audit Commission. (1998). *Ghost in the machine: An analysis of IT fraud and abuse.* Audit Commission Publications.

Audit Commission. (2001). *yourbusiness@risk: An update on IT abuse 2001.* Audit Commission Publications.

Brown, L. (2000). Lifelong learning: Ideas and achievements at the threshold of the twenty-first century. *Compare: A Journal of Comparative and International Education, 30*(3), 341–351.

Hoffman, D. L., & Novak, T. P. (1998). Bridging the digital divide: The impact of race on computer access and internet use. *Science, 280*(4), 390–391.

Jisc. (2003). Introducing managed learning environments (MLEs). *Jisc.* Retrieved July 7, 2005, from www.jisc.ac.uk/index.cfm?name=mle_briefingpack.

Kanes, C. (2009). Challenging professionalism. In C. Kanes (Ed.), *Elaborating professionalism* (pp. 1–15). Springer. https://doi.org/10.1007/978-90-481-2605-7_1

Kemmis, S. (2009). What is professional practice? Recognising and respecting diversity in understandings of practice. In C. Kanes (Ed.), *Elaborating professionalism* (pp. 139–165). Springer. https://doi.org/10.1007/978-90-481-2605-7_8

Maner, W. (1996). Unique ethical problems in information technology. *Science and Engineering Ethics, 2*(2), 137–154.

Marcuse, H. (1964). *One-dimensional man: Studies in the ideology of advanced industrial society.* Beacon.

Rogerson, S. (2021). *The evolving landscape of ethical digital technology.* Taylor & Francis Group, ISBN 9781003203032.

Toffler, A. (1970). *Future shock.* Bantam Books.

Trotter, W. (2005). *Asia Pacific Forum – Stretch your horizons.* Retrieved July 7, 2005, from www.global-learning.de/g-learn/cgibin/gl_userpage.cgiStructuredContent=m130339.

Whitworth, A. (2005). The politics of virtual learning environments: Environmental change, conflict, and e-learning. *British Journal of Educational Technology, 36*(4), 685–691.

Chapter 5

Regulation and Policy

Regulations and policies sit beneath governance which is about providing, distributing and consequently regulating.

> Governance must address both the process and product dimensions of digital technology . . . It must promote a sense of obligation in professional developers, thus ensuring that digital technology products and services are fit-for-purpose. Governance implies a system in which all stakeholders have the necessary input into the decision-making process. . . . Governance should have its foundation as delivering ethical, efficient and effective digital technology. These three factors must be multiplicative rather than summative.
>
> (Rogerson, 2021, p. 301) (This is discussed further in Section 5.12.)

Regulations form the subset of governance which focuses on the flow of events and behaviour. They are designed to make people comply and behave in a certain manner. Policies are rules made by organisations to achieve their aims, objectives and goals. Matthews, Rice, and Quan (2021, p. 55) explain that,

> Policy and regulation tend to be associated exclusively with government action, though importantly, they exist elsewhere in other forms. Just as governments establish policy and regulation in their respective jurisdictions, market actors like professional bodies and industry associations establish professional standards, codes of conduct, industry certifications, and more. Together, government and market-led responses can seek to govern innovation, mitigate negative impacts of technology, and maximize positive ones.

They (ibid, p. 58) point out that, "best practices for policy and regulation are informed by discussion with academic, civil society, and industry experts, whose work focuses in detail on the social impacts of technology."

During the period covered by this book data protection has been at the forefront of regulatory change. However, many other areas have been addressed through regulation and institutional policy. This wide landscape is the focus of this chapter.

DOI: 10.1201/9781003309079-5

5.1 Data Matching [1997]*

This section was first published as: Rogerson, S. (1997) ETHIcol—Data Matching. IMIS Journal, Vol 7 No 1, p. 21. Copyright © Simon Rogerson.

Whilst fraud is accepted by society as being wrong, the methods and in particular data matching deployed to prevent or detect fraud are open to criticisms of invasion of privacy and restricting freedom of the individual. There is a difficult balance to be struck between the rights of individuals and the devastating financial losses particularly of public funds. It is unclear who should judge whether the cause legitimises the data matching and data sharing activity and how that judgement should be derived.

Data matching is the computerised comparison of two or more sets of records which relate to the same individual. It is primarily used as a method for combating fraud. There is increasing use of data matching by both public and private organisations in an attempt to reduce fraudulent activity which has been estimated to run annually into billions of pounds in the UK. The relative cheapness and availability of sophisticated processing means that data matching is likely to increase even more rapidly.

There are a number of examples of data matching being undertaken by government agencies. The DSS has established a Housing Benefit Matching Service aimed at detecting benefit fraud. The Audit Commission uses data matching across local authorities regarding benefit claimants, education awards and activities of local authority employees. The Social Security Administration (Fraud) Bill provides for wider sharing by central and local government and the Post Office for fraud prevention or detection purposes.

There is a difference between the methods of detection used in the past and data matching. Traditional investigation is triggered by some evidence of a wrongdoing by an individual, such as tax evasion or bogus benefit claims. Data matching is not targeted at individuals but at entire categories of people. It is initiated not by the suspicion concerning an individual but because the profile of a particular group is of interest. This leads to three issues of concern.

Privacy—Data matching is likely to involve matching personal records compiled for unrelated purposes. Surely an individual has a right to control personal information and prevent its use without consent for purposes unrelated to those for which it was collected.

Due process of the law—Once a match has been undertaken it will result in a number of hits. All those identified are in jeopardy of being found guilty of a wrongdoing. It is unlikely that these individuals are given any notice of their situation, since doing so might affect the investigation, or an opportunity to contest the results of the match at an early stage. For these reasons their right to due process of law is curtailed.

Presumption of innocence—The presumption of innocence is intended to protect people against having to prove that they are free from guilt whenever they are investigated. Data matching can reverse this to a presumption of guilt. This is because the technology of data matching is so plausible and the detection of fraud is much applauded. These powerful influences will weigh heavy in favour of the notion that those identified must be guilty.

One of the advances in data matching techniques is the automatic sharing of information. This is dependent on increasingly complex and expanding communications networks that link more and more organisations together. There is a tendency to view people as objects to be moved about this network. In this instance it is unclear how an innocent citizen might be assured that incorrect information is totally corrected across the complex web of organisational relationships. The security of such a network is also of concern as the network will be as secure as its least secure node. Individual organisations might not be aware of all other organisations linked to the network for if they were they might be wary of sharing information due to insecurity through either technological deficiencies or moral attitudes.

In order to detect sophisticated fraud there is need to use complex data analysis techniques which may well involve methods based on partial match interpretation which in turn increases the risks of incorrect hits. Simple fraud detection lends itself to data matching systems that have little or no human intervention and the pressure to use such systems will grow. There needs to a be clear understanding of who has access to this information and how it will be used. Legislative frameworks, codes of practice and operational procedures must be in place to ensure that data matching is applied in a balanced way sensitive to the needs of organisations and society as a whole and to the rights of individuals.

Each of us is probably the subject of some data matching exercise quite frequently. It might be related to a mortgage application or a latest tax return or many other reasons. How do you feel about that? How would you feel if you were on the receiving end of an erroneous data match which led to a mortgage foreclosure or the loss of your credit rating? Next time you are involved in implementing a data matching system think on it could be you incorrectly caught by it!

5.2 Privacy [1997]*

*This section was first published as: Rogerson, S. (1997) ETHIcol—Privacy. *IMIS Journal*, Vol 7 No 3, p. 21. Copyright © Simon Rogerson.

Within industry and commerce privacy continues to be an important issue and one which regularly presents tensions between competing obligations. Privacy is a fundamental right of individuals and is an essential condition for the exercise in self-determination. At one level privacy is about corporate ethos and how individual rights are valued. This leads to the consideration of what levels of protection should be in place. This in turn leads to the lowest level which is concerned with the tools that might be used to manage privacy such as encryption, codes of practice and role definitions for staff.

Balancing the rights and interests of different parties in a free society is difficult. The acceptable balance will be specific to the context of a particular relationship and will be dependent upon trust between concerned parties and subscription to the principle of informed consent. This balance might incur the problem of protecting individual privacy while satisfying government and business needs. Such problems are indicative of a society that is becoming increasingly technologically dependent.

Sometimes individuals must give up some of their personal privacy in order to achieve some overall social benefit. For example, a social services department might hold sensitive information about individuals that provides an accurate profile of individual tendencies, convictions and so on. The sharing of this data with, for example, the local education authority in cases of child sex offenders living in the area might be considered morally justified even though it might breach individual privacy.

Organisations are increasingly computerising the processing of personal information. This may be without the consent or knowledge of the individuals concerned. There has been a growth in databases holding personal and other sensitive information in multiple formats of text, pictures and sound. This data must be accurate and objective. The scale and type of data collected and the scale and speed of data exchange have changed with the advent of computers. The potential to breach people's privacy at less cost and to greater advantage continues to increase.

Computer privacy is a new twist on an old ethical problem and involves issues not previously raised or cannot be predicted. For example, advances in genetic data have led to some interesting ethical questions as it can accurately define genetic relatives and thus establish hereditary traits and diseases. Individuals have certain rights to how and where that information is distributed but in order to exercise those rights they will undoubtedly learn of their genetic profiles and there is the new twist. Knowledge of one's genetic profile will undoubtedly affect the individual's self-perception, self-esteem, and lifestyle. Thus, privacy in this situation must include an individual's right not to know.

Within industry and commerce there are two important categories of privacy: consumer privacy and employee privacy. Consumer privacy considers the information complied by data collectors such as marketing firms, insurance companies and retailers, the use of credit information collected by credit agencies and the rights of the consumers to control information about themselves and their commercial transactions. Indeed, the extensive sharing of personal data is an erosion of privacy that reduces the capacity of individuals to retain control of factors which may affect their lives. Organisations involved in such activities have a responsibility to ensure privacy rights are upheld. The problems involved in the transfer of consumer data include:

- the potential for data to be sold to unscrupulous vendors
- the problems with ensuring the trustworthiness and care of data collectors
- the potential for combining data in new and novel ways to create detailed composite profiles of individuals or categories of people
- the difficulty of correcting inaccurate information once it has been distributed across many different files

Employee privacy deals primarily with the growing reliance on electronic monitoring and other mechanisms to analyse work habits and measure employee productivity. Employees have privacy rights including:

- the right to control or limit access to personal information provided to an employer
- the right to choose what he or she does outside the workplace
- the right to privacy of thought
- the right to autonomy and freedom of expression

In the modern workplace there are increasing opportunities to monitor activity. It is important to ensure the use of monitoring facilities does not violate the privacy rights of employees. Some of the potential problem areas are:

- network management programs of personal computers that allow user files and directories to be monitored and to track what is being typed on individual computer screens
- network management systems that enable interception and scrutiny of communications among different offices and between remote locations
- email systems that generate archives of messages that can be inspected by anyone with the authority or technical ability to do so
- broad-based electronic monitoring programs that track worker productivity and work habits
- close circuit television surveillance systems that are computer controlled, have extensive archiving facilities and digital matching facilities

Privacy legislation is not universal and where it exists in some countries it is often ineffective. This is because the legislative time frame is always much slower that the technological one.

Furthermore, given communities are becoming increasingly free of geographical and temporal constraints the viability of privacy legislation is questionable. Resolution of privacy problems is thus reliant upon organisations fulfilling their obligations as a supplier, a client, an end user of IT and a community member and being committed to self-regulation.

Responsible organisations will ensure that the privacy of the individual is protected whilst they pursue their business activities. The challenging question is do you think your organisation responsibly addresses privacy and if not what are you going to do about it?

5.3 Corporate Issues [1998]*

*This section was first published as: Rogerson, S. (1998) ETHIcol—Corporate Issues. IMIS Journal, Vol 8 No 2, pp. 6–7. Copyright © Simon Rogerson.

Organisations are continuing to invest heavily in computer systems designed to improve efficiency and effectiveness. These new systems are often complex replacing manual processes that previously defied automation, or are second or third generation systems, introducing new functions into already automated processes. Usually new systems are not simply a matter of giving staff a better tool to do the same work but involve changes to the nature of the work itself. This in turn changes the very fabric of the society in which we live. Careful planning and consultation are needed to implement new systems successfully. All those affected by the change must be involved in good time in an appropriate way. They may include customers, suppliers, regulators, business partners and members of the public as well as employees.

The impact of new systems will usually be judged in terms of whether the gains in efficiency and effectiveness are realised as planned, but that is not all. The new systems may involve changes in the staffing levels, organisational structure, and social groupings. These changes can affect staff morale inside an organisation and relationships outside, particularly with customers, to such an extent that they may cause disadvantages to the organisation which reduce, or even outweigh, the basic benefits achieved. There is a growing realisation within the UK that good ethics is good business. The latest survey by the Institute of Business Ethics shows a dramatic increase in the number of organisations that have a corporate code of ethics.

Given the central and essential role of Information and Communication Technologies (ICT) in organisations it is paramount that this ethical sensitivity percolates decisions and activities related to ICT. In particular organisations need to consider:

■ how to set up a strategic framework for ICT that recognises personal and corporate ethical issues
■ how the methods for systems development balance ethical, economic and technological considerations
■ the intellectual property issues surrounding software and data
■ the way information has become a key resource for organisations and how to safeguard the integrity of this information
■ the increasing organisational responsibility to ensure that privacy rights are not violated as more information about individuals is held electronically
■ the growing opportunity to misuse ICT given the increasing dependence of organisations on it and the organisational duty to minimise this opportunity whilst accepting individuals have a responsibility to resist it
■ the way advances in ICT can cause organisations to change their form—the full impact of such change needs to be considered and, if possible, in advance, and the way the advent

of the global Information Society raises new issues for organisations in how they operate, compete, co-operate and obey legislation

■ how to cope with the enormous and rapid change in ICT, and how to recognise and address the ethical issues that each advance brings

Thus, there is an ethical agenda associated with the use of ICT in organisations. This agenda combines issues common to many professions and issues that are specific to ICT. New advances in ICT and new applications may change the agenda. If organisations wish to secure benefits to their business in the long term and enhance their reputation they have to address a comprehensive agenda. The following steps provide a way in which organisations can establish such an agenda and address the ethical issues arising in the field of ICT.

1. Decide the organisation's policy, in broad terms, in relation to ICT. This should:
   ■ take account of the overall objectives of the organisation, drawing from such existing sources as the organisational plan or mission statement
   ■ use the organisation's established values, possibly set out in its code of practice, for guidance in determining how to resolve ethical issues
   ■ set the scope of policy in terms of matters to be covered
2. Form a statement of principles related to ICT that would probably include:
   ■ respect for privacy and confidentiality
   ■ avoid ICT misuse
   ■ avoid ambiguity regarding ICT status, use, and capability
   ■ be committed to transparency of actions and decisions related to ICT
   ■ adhere to relevant laws and observe the spirit of such laws
   ■ support and promote the definition of standards in, for example, development, documentation and training
   ■ abide by relevant professional codes
3. Identify the key areas where ethical issues may arise for the organisation, such as:
   ■ ownership of software and data
   ■ integrity of data
   ■ preservation of privacy
   ■ prevention of fraud and computer misuse
   ■ the creation and retention of documentation
   ■ the effect of change on people both employees and others
   ■ global ICT
4. Consider the application of policy and determine in detail the approach to each area of sensitivity that has been identified.
5. Communicate practical guidance to all employees, covering:
   ■ the clear definition and assignment of responsibilities
   ■ awareness training on ethical sensitivities
   ■ the legal position regarding intellectual property, data protection and privacy
   ■ the explicit consideration of social cost and benefit of ICT application
   ■ the testing of systems (including risk assessment where public health, safety and welfare, or environmental concerns arise)
   ■ documentation standards
   ■ security and data protection

6. Whilst organisations have a responsibility to act ethically in the use of ICT so to do individual employees. Those involved in providing ICT facilities should support the ethical agenda of the organisation and in the course of their work should:
   ■ consider broadly who is affected by their work
   ■ examine if others are being treated with respect
   ■ consider how the public would view their decisions and actions
   ■ analyse how the least empowered will be affected by their decisions and actions
   ■ consider if their decisions and acts are worthy of the model ICT professional

Note: This article is based upon extracts from *Ethical Aspects of Information Technology: Issues for Senior Executives* by Simon Rogerson and published by the Institute of Business Ethics in 1998

5.4 Data Protection for the People [1999]*

*This section was first published as: Rogerson, S. (1999) ETHIcol—Data Protection for the People, IMIS Journal, Vol 9 No 6, pp. 24–25. Copyright © Simon Rogerson.

It is a prerequisite that people within a civilised society should be law abiding. Indeed, from the earliest times laws have been created to sustain and develop society so that its citizens can live peaceably, and to provide some form of address against those who go astray. It is tempting and often the case that the role of law is interpreted as simply being a control mechanism rather than providing guidance as to what is considered as being acceptable behaviour. This leads to an attitude of legal compliance rather than seeing the law as a minimum acceptable standard. Those involved in undertaking or controlling computer-related activity regularly adopt a legal compliance attitude. This however may not be enough because, in all judgements relating to the development and application of information and communication technology, concern for the health, safety, and welfare of the public must be primary. Privacy of the individual is a good illustration of this point.

The 1984 Data Protection Act in the UK did not really address the issue of privacy of the individual because, for example, it did not cover all data processing and storage, it had many exemptions explicitly specified and it was driven by political concern over the economic repercussions of companies leaving the UK to establish their operations elsewhere in Europe if an act, which was compliant with the European Directive, was not in place. Primarily the Act required organisations to register their computerised data processing activities of personal data. Legal compliance did very little to sustain, let alone improve, an individual's right to privacy.

The new Data Protection Act received Royal Assent on the 16 July 1998. The Act and the secondary legislation required to support it, will be brought into force on the 1 March 2000. The new Act has a clear focus on individual privacy. The Data Protection Registrar's initial interpretation of the new Act illustrates this. The new Act expressly provides that personal data are not to be treated as processed fairly unless, as far as practicable, the following criteria are met:

■ the individual has given his or her consent to the processing
■ the processing is necessary for the performance of a contract with the individual
■ the processing is required under a legal obligation

- the processing is necessary to protect the vital interests of the individual or to carry out public functions
- the processing is necessary in order to pursue the legitimate interests of the data controller or certain third parties (unless prejudicial to the interests of the individual)

Stricter conditions apply to the processing of sensitive data. This category includes information relating to racial or ethnic origin, political opinions, religious or other beliefs, trade union membership, health, sex life, and criminal convictions.

There is a new eighth principle restricting the transfer of personal data outside the EU. Personal data may only be transferred to third countries if those countries ensure an adequate level of protection for the rights and freedoms of data subjects. When determining adequacy data controllers should consider, for example, the nature of the data, the country of origin and final destination, and the law or any relevant codes of conduct in force.

The definition of data in the new Act has been extended so that it now catches information which is recorded as part of a relevant filing system where the records are structured, either by reference to individuals or by reference to criteria relating to individuals, so that specific information relating to a particular individual is readily accessible. This definition will catch some types of manual data as well as automated forms of processing.

Clearly, the new Act has much to offer regarding the safeguarding of personal privacy but does it go far enough? Consider the case of health data. In its recently published opinion on the ethical issues of healthcare in the Information Society, the European Group on Ethics in Science and New Technologies (EGE) explained that personal health data necessarily touch upon the identity and private life of the individual and are thus extremely sensitive and that confidentiality of personal health data must be guaranteed at all times. This implies that the informed consent of the individual is required for the collection and release of such data.

The new data protection legislation upholds much of this requirement but there are shortcomings. Probably the most notable relates to EGE's comment that the respect for the confidentiality of health data continues after the death of the person. Once a person dies he or she is no longer a data subject and therefore the Act no longer applies which means the personal health data can be distributed, processed, and published without restriction under the Act. This seems unacceptable as personal health data of dead people may well have repercussions for those still living. Responsible organisations and individuals handling such data are likely to respect this implication and consequently will still maintain data confidentiality. However, the unscrupulous might see an opportunity to benefit from this situation by making once restricted data available in return for a fee. Many would argue this might be legal but it is not ethical.

The new Act offers a great opportunity for computer practitioners at all levels to improve their level of professionalism regarding the public. The Act is a clear indication that respect of an individual's privacy is paramount. Why not use the opportunity to review the processing of personal data and implement what is right for the individual rather than simply achieving legal compliance? Put yourself in the place of the data subject and ask yourself if you are happy with the amount of data that is collected, stored, manipulated and transmitted about you in the systems that you are running regardless of the fact that they are legally compliant. If the answer is no then do something about it!

5.5 Serving Two Masters [2001]*

This section was first published as: Rogerson, S. (2001) ETHIcol—Serving Two Masters. IMIS Journal, Vol 10 No 6, pp. 28–29. Copyright © Simon Rogerson.

Readers in the UK will be aware of the Human Rights Act which recently came into force. This act incorporates the European Convention on Human Rights. Quite rightly it has made our responsibilities to others high profile. So what does this mean to ICT professionals?

The Act lays out several rights, many of which have a bearing on what activities ICT professionals undertake and how they perform those activities. The most relevant rights appear to be:

- everyone has the right to liberty and security of person
- everyone has the right to respect for his private and family life, his home and his correspondence
- everyone has the right to freedom of thought, conscience and religion
- everyone has the right to freedom of expression
- everyone has the right to freedom of peaceful assembly and to freedom of association with others
- rights and freedoms shall be secured without discrimination on any grounds

According to Ros Taylor writing in the Guardian on 29 March 2000, the Act makes public services, including the police, schools, local government and hospitals, potentially liable for breaches of the Act. She reports that many organisations that consider themselves outside the public sector could also be liable. These include private schools, hospitals, prisons and nursing homes, because they carry out public sector duties. Clearly there are many ICT professionals working for such organisations and they must review how they treat the public and make sure they operate in a fair, reasonable, and legal way. ICT professionals provide the systems that process the data that produces the information on which decisions are based. In practice it is the ICT professional who is best placed to ensure such systems do not contravene the Act. There can however be pressures to act in a different way.

At the time of writing this column another piece of legislation came into force, the Regulation of Investigatory Powers (RIP) Act. According to Will Knight and Wendy McAuliffe of ZDNet (UK), "The RIP Act gives law enforcers power to intercept communications via devices to be installed at ISPs and to imprison those who fail to hand over the keys to encrypted messages." They explain that it also allows "employers to spy on the internet activities of staff without their consent." ICT professionals will have a fundamental role to play in implementing computer systems which provide the means under the powers of the RIP Act to investigate the activities of individuals and groups.

Whilst there may be justifiable reasons to monitor activity there are most certainly many reasons that cannot be justified. Indeed, David Kilduff, Partner and Head of Public Sector and Private Finance of Walker Morris explains that the Human Rights Act is likely to have important repercussions on the way organisations operate. He points out that, "Surveillance and security systems (including computer sweeping) and data capture or management policies and processes will need to be reviewed to ensure that the right of privacy is not compromised." If you are in the position of developing a monitoring system then consider the list of rights outline previously and ask yourself whether any of these rights appear to be contravened. If the answer is yes then clearly there needs to be a thorough review of the reasons behind the need to monitor.

There are many who believe that the RIP Act contradicts the objectives of the Human Rights Act. If so then the ICT professional is left in a precarious position as legal compliance with both acts may not be feasible. This is an occasion when the professional must decide which is the course of action to take that promotes social responsibility.

As for the legal position, Knight and McAuliffe point out that, "Thanks to the Human Rights Act it will be possible for the RIP to be challenged in a UK court by someone who feels that their right to privacy has been denied. It will then be up to a judge to decide whether there is conflict."

The UK Government has established the Human Rights Unit at the Home Office. The slogan of the Home Office is "building a safe, just and tolerant society." This seems like a good stance for ICT professionals to take in this uncertain climate. What do you think?

5.6 New Code of Ethics [2001]*

*This section was first published as: Rogerson, S., McRobb, S & Fairweather, N.B. (2001) ETHIcol— New Code of Ethics. IMIS Journal, Vol 11 No 1, pp. 17–19. Copyright © Simon Rogerson, Steve McRobb & N Ben Fairweather.

In the April 2000 issue of the IMIS Journal members were encouraged to comment on a draft Code of Ethics for the Institute by completing a survey form. A total of 73 feedback forms were returned. This may seem a relatively small proportion of the total membership of the Institute, but it is typical for a survey of this kind. For example, a similar consultation exercise on a draft of the ACM/IEEE Code of Ethics generated a comparable level of response, expressed as a percentage of total membership of the ACM and IEEE combined.

Each respondent could cast 40 votes, one for each detailed provision in the code. On this basis, 2608 votes were cast in total (not quite the expected number as a few respondents didn't answer every question). Of these, a total of 1303 votes (exactly 50%) strongly agreed with the corresponding provision and a further 1080 votes (41.4 %) agreed (so 2383–91.4% agreed or strongly agreed). Only 163 votes (6.3%) were uncertain, 46 votes (1.8%) opposed and 16 votes (0.6%) strongly opposed (so 62–2.4% opposed or strongly opposed). It is perhaps also worth noting that two individuals recorded 10 of the 16 votes of strong opposition.

In addition to signalling their support for, or opposition to, the various provisions of the draft code, a significant number of members also took the opportunity to provide supplementary comments. These comments fall into three broad groups, each of value in a different way.

Some of the comments either address a specific provision in the code or make a more general point about what should, or should not, be included within the code. Many of this type have been very helpful in reviewing the code, and indeed some revisions have been made specifically in response to concerns expressed by members in this way. For example, a number of comments were aimed at explaining a member's opposition to a particular provision. In some cases this highlighted a weakness in logic or wording, and these have been addressed as far as possible. Of course, it has not been possible, nor would it be appropriate, to meet all the concerns expressed. For example, some members are simply opposed to a principle that is strongly supported by the great majority.

The second broad group of comments is interesting, not because they suggest any particular amendment to the code, but rather because they serve to highlight areas of ambiguity or conflicts between different provisions. To some extent the issues that have triggered these comments can also be discerned in the pattern of voting on some provisions. To give one striking example, the code's provision number 4.5 embodies a general opposition to covert surveillance. This proved contentious, not because there was significant opposition to it in principle, but because a number of members believe that covert surveillance is justifiable in certain circumstances, for example, where criminal activity is suspected. Some expressed this view explicitly in the form of a verbal comment while still voting in favour of the draft provision. Others indicated their uncertainty over this provision with their vote but made no supplementary comment. It appears likely that these two responses can be seen as different expressions of a similar ethical uncertainty. We would argue that the wording of provision 4.5 allows in a reasonable way for the ethical ambiguities of this provision: *I will actively oppose surveillance undertaken without informed consent of the subjects, unless such surveillance is justified by a greater ethical priority.* In any case, both votes and comments reinforce the conclusion that this issue is something of an ethical hotspot.

The third general group of comments relates not to the content of the code itself but to the consequences of its adoption. Typical concerns here relate to the perceived need for other supporting structures and activities to be developed if the code is to have a real impact on the Institute and its members. To give an example, several members expressed a worry that in certain circumstances the code could be seen as asking a professional to place his or her job, career and even personal liberty on the line. For example, provision 1.2 would (under certain conditions) encourage whistleblowing, and provision 2.9 would (under certain conditions) encourage law breaking.

What follows is the revised code which takes into account both the votes cast and comments made.

5.6.1 Preamble to Code

The Institute for the Management of Information Systems has a vision to see Information Systems Management regarded as one of the key professions influencing the future of our society. Along with that recognition, however, comes a responsibility for practitioners to adhere to professional level standards of training and codes of conduct.

This Code of Ethics details an ethical basis for the practitioner's professional commitment. It does this by summarising the ethical values that the Institute expects all members to uphold and the ethical standards that a member should strive to achieve. These values and standards should guide the professional conduct of a member at all times.

In common with other Codes of Ethics, the Code is meant to be taken holistically—the conscientious professional should take account of all principles and clauses that have a bearing on a given set of circumstances before reaching a judgement on how to act. Selected parts of the Code should not be used in isolation to justify an action or inaction. Nor should the absence of direct guidance in the Code on a specific issue be seen as excusing a failure to consider the ethical dimensions of an action or inaction.

It is neither desirable nor possible for a Code of Ethics to act as a set of algorithmic rules that, if followed scrupulously, will lead to ethical behaviour at all times in all situations. There are likely to be times when different parts of the code will conflict with each other. There may also be times when parts of this code will conflict with other ethical codes or generally accepted priorities in the wider world. At such times, the professional should reflect on the principles and the underlying spirit of the Code and strive to achieve a balance that is most in harmony with the aims of the Code. Some indication of relative priority is given within the code where conflict can be anticipated. However, in cases where it is not possible to reconcile the guidance given by different articles of the code, the public good shall at all times be held paramount.

5.6.2 Fundamental Principles

Every Fellow and Member of the Institute (including both Professional and Affiliate Membership grades) shall employ all his or her intelligence, skills, power and position to ensure that the contribution made by the profession to society is both beneficial and respected. In accordance with this commitment, he or she shall at all times uphold the following six fundamental principles:

Principle 1: Society—I will uphold the health, safety and welfare of wider society, future generations and the environment.

Principle 2: Organisations—I will serve my employers and clients honestly, competently and diligently.

Principle 3: Peers—I will respect and support the legitimate needs, interests and aspirations of all my colleagues and peers.

Principle 4: Staff—I will encourage and assist those I supervise both to fulfil their responsibilities and to develop their full potential.

Principle 5: Profession—I will strive to be a fit representative of my profession and to promote the vision of the Institute.

Principle 6: Self—I will be honest in representing myself and will continually strive to enhance both my professional competence and my ethical understanding.

5.6.3 The Code in Detail

1. **Society**: I will uphold the health, safety, and welfare of wider society, future generations and the environment.

   1.1 I will strive to ensure that those professional activities for which I have responsibility, or over which I have influence, will not be a cause of avoidable harm to any section of the wider community, present or future, or to the environment.

   1.2 When there is no effective alternative I will bring to the attention of the relevant public authorities any activity by staff I supervise, colleagues, employers, clients or fellow professionals that is likely to result in harm as described under article 1.1.

   1.3 I will contribute to public debate regarding policy formulation in areas where this is in the wider interest, I have technical or professional competence and there is an appropriate opportunity to do so.

   1.4 I will use my knowledge, understanding, and position to oppose false claims made by others regarding the capabilities, potential or safety of any aspect of Information Systems and Information or Communication Technology.

   1.5 I will strive to protect the legitimate privacy and property of individuals and organisations in wider society, where there is a risk that these may be compromised by professional activities for which I am responsible, or over which I have influence.

2. **Organisations**: I will serve my employers and clients honestly, competently and diligently.

   2.1 I will endeavour to avoid, identify, and resolve conflicts of interest.

   2.2 I will accept neither an assignment that I know I will not be able to complete competently, nor an assignment that I suspect I will not be able to complete competently unless the risks are knowingly and freely accepted by all parties concerned.

   2.3 I will not knowingly commit a team to a task that cannot be completed within acceptable limits of cost, effort, and time, unless the risks are knowingly and freely accepted by all parties concerned.

   2.4 I will preserve the legitimate confidentiality of the affairs of my employers and clients.

   2.5 I will protect the legitimate property and uphold the legitimate rights of my employers and clients.

   2.6 I will adhere to relevant and well-founded organisational and professional policies and standards.

   2.7 I will ensure, within the extent of my influence, that sufficient and competent staff are deployed on any professional activity.

   2.8 I will ensure, within the extent of my influence, compliance with relevant and well-founded technical standards and methods.

2.9 I will ensure that I do not cause my employers or clients to breach applicable legislation or well-founded rules, unless there is a greater ethical priority of sufficient magnitude.

3. **Peers**: I will respect and support the legitimate needs, interests and aspirations of my colleagues and peers.

3.1 I will protect the legitimate privacy and property of my colleagues and peers.

3.2 I will refrain from all conduct that inappropriately undermines my colleagues or peers.

3.3 I will give an honest opinion regarding the competence and potential of my colleagues and peers, when it is appropriate to do so.

3.4 I will act in support of colleagues and peers who uphold what is right above their personal benefit and convenience.

3.5 I will promote teamwork among my colleagues and peers, taking my fair share of the burdens and no more than my fair share of the credit.

4. **Staff**: I will respect and support the legitimate needs, interests and aspirations of those I supervise and I will encourage and assist them both to fulfil their responsibilities and to develop their career potential.

4.1 I will adopt and promote an ethical approach to management.

4.2 I will be fair in my dealings with those I supervise.

4.3 I will be open towards those I supervise, unless constrained by a greater ethical priority.

4.4 I will actively oppose discrimination at work except on the sole basis of an individual's capacity for the task, and will take care that my judgement on this issue is not prejudiced by preconceived notions regarding any group in society.

4.5 I will actively oppose surveillance undertaken without informed consent of the subjects, unless such surveillance is justified by a greater ethical priority.

4.6 I will encourage staff education, training, development, and promotion, and will represent the legitimate best interests of those I supervise in developing their careers both within and beyond the organisation.

4.7 I will give an honest opinion regarding the competence and potential of staff I supervise, when it is appropriate to do so.

4.8 I will protect the legitimate privacy and property of those I supervise.

4.9 I will promote adherence to relevant and well-founded specialist codes of conduct.

4.10 I will promote teamwork among those I supervise, taking my fair share of the burdens and no more than my fair share of the credit.

4.11 I will not require those I supervise to breach applicable legislation or well-founded rules.

5. **Profession**: I will strive to be a fit representative of my profession and to promote the vision of the Institute.

5.1 I will act with integrity at all times.

5.2 I will be honest unless constrained by a greater ethical priority.

5.3 I will strive to abide by this Code of Ethics and thereby enhance the public image and standing of the profession.

5.4 I will be willing to perform voluntary work on behalf of the profession, provided that I have the necessary time, resources, and capability for the task.

6. **Self**: I will be honest in representing myself and will continually strive to enhance both my professional competence and my ethical understanding.

6.1 I will maintain my personal integrity.

6.2 I will not allow personal interests to influence the advice I give on technical and professional matters.

6.3 I will maintain the continuing development of my technical, professional and ethical understanding, and competence.

5.7 Police Intelligence? [2004]*

This section was first published as: Rogerson, S. (2004) ETHIcol—Police Intelligence? IMIS Journal, Vol 14 No 4, pp. 31–32. Copyright © Simon Rogerson.

In 2001 Ian Huntley was appointed to Soham Village College. This led to the horrendous murders of Jessica Chapman and Holly Wells for which Huntley was convicted on 17 December 2003. The heinous crime and the unimaginable grief for the families of the girls shocked the nation. It became clear that Huntley had been known by the police since 1995 in relation to various sexual offences. It defied belief that this information failed to come to light during the appointment process in 2001. This revelation ultimately led to the Bichard Inquiry into child protection procedures in Humberside police and Cambridgeshire Constabulary, particularly the effectiveness of relevant intelligence-based record keeping, vetting practices since 1995 and information sharing with other agencies.

The Bichard Report was published on 22 June 2004. This edition of ETHIcol looks at some of the information systems issues raised by the report (the numbers in brackets refer to paragraphs or recommendations in the report).

According to the report the current situation across the 43 police forces in England and Wales is not good. There is no common IT system for managing criminal intelligence (3.61). Indeed, this strategic objective was removed from the national police IT strategy in 2000 as it was unachievable because of lack of funds. Without a strong implemented national strategy it is hardly surprising that police forces cannot share intelligence. Indeed, Bichard found (3.63) that each police force has a variety of IT systems that are being used for many different purposes. Interfaces between systems at force-to-force level are almost non-existent. It is incredible that it was discovered that even within forces system interaction has been patchy at best.

Bichard found contradictory evidence regarding information management. "The National Intelligence Model (NIM) [exists as] a management framework that requires police forces to analyse and address the methods by which intelligence is obtained, created, stored, and used. Its aim is to enhance intelligence-led policing." (3.64) However, it was discovered that it had done little in reality. There remains "a lack of clear, national guidance for the police about information management—the way in which information is recorded[,] reviewed, retained or deleted" (3.66). It was found that each police force had its local interpretation of NIM (3.81) which clearly makes intelligence sharing virtually impossible.

Of the 31 recommendations made in the report, 11 related directly to IT. Many others had implications for IT in that effective vetting was not possible without computerised criminal intelligence support that worked. The recommendations recognise that attempts had been made to improve the situation but there is implication of these being too little, too late and too slow. The 11 specific recommendations were:

- A national IT system for England and Wales to support police intelligence should be introduced as a matter of urgency (1).
- The pilot PLX system, which flags that intelligence is held about someone by particular police forces, should be introduced in England and Wales by 2005 (2).
- The procurement of IT systems by the police should be reviewed to ensure that, wherever possible, national solutions are delivered to national problems (3).
- Investment should be made available by Government to secure the Police National Computer's (PNC) medium and long-term future, given its importance to intelligence-led policing and to the criminal justice system as a whole (4).

- The planned new Code of Practice, made under the Police Reform Act 2002, dealing with the quality and timeliness of PNC data input, should be implemented as soon as possible (5).
- The quality and timeliness of PNC data input should be routinely inspected (6).
- The transfer of responsibility for inputting court results onto the PNC should be reaffirmed by the Court Service and the Home Office and, if possible, accelerated ahead of the 2006 target (7).
- A Code of Practice should be produced covering record creation, review, retention, deletion and information sharing. This needs to be clear, concise and practical. It should supersede existing guidance (8).
- The Code of Practice must clearly set out the key principles of good information management (capture, review, retention, deletion and sharing), having regard to policing purposes, the rights of the individual and the law (9).
- The Code of Practice must set out the standards to be met in terms of systems (including IT), accountability, training, resources and audit. These standards should be capable of being monitored (10).
- The Code of Practice should have particular regard to the factors to be considered when reviewing the retention or deletion of intelligence in cases of sexual offences (11).

The obviousness of these recommendations is frightening. How could good practice and common sense be so lacking in such an important area of computing? What sort of input and influence did IS professionals have in the strategic development and implementation of computerised criminal intelligence support? If they had concerns did they raise them and were they ignored? It is interesting that Bichard does not address the role and influence of the IS professionals in this situation. It seems this is a crucial issue.

If ever there was a poignant example of the need for joint decision making and action by strategists and technologists within the information systems field this is it. We cannot say that relevant and timely information provided to those appointing Huntley would have prevented the murders but it would have reduced the likelihood of opportunity. For information systems professionals it is a salutary reminder of our obligations to society and our responsibility to provide systems that sustain and promote public good.

5.8 Justice for All [2006]*

This section was first published as: Rogerson, S. (2006) ETHIcol—Justice for All. IMIS Journal, Vol 16 No 1, pp. 31–32. Copyright © Simon Rogerson.

The UK Government recently published "Transformational Government Enabled by Technology" in which it laid out its "strategy to seize the opportunity provided by technology to transform the business of government." Claiming that, "Technology has a major part to play in the solutions to each of three major challenges which globalisation is setting modern governments—economic productivity, social justice and public service reform." One of the key drivers for this report was the recognition that, "The capacity and capability of (particularly central) government organisations and their suppliers to deliver technology-enabled business change has been subject to severe criticism by Parliament and the press over the last decade. Public confidence in government's ability to deliver technology projects reached a low point by the late 1990s."

Social justice is the ideal condition in which all members of a society have the same basic rights, security, opportunities, obligations and social benefits. The government publication contains many statements which directly and indirectly relate to social justice. Some are now considered.

"Overall this technology-enabled transformation will help ensure that . . . citizens feel more engaged with the processes of democratic government."—Current evidence does not seem to support this view. The postponement of electronic voting in the UK was because of lack of public confidence that this would deliver benefit.

"There are new information assurance risks: terrorists, organised criminals and hackers threaten information and services, and theft of identity and of personal data is of increasing concern to individuals and businesses."—Government services are particularly vulnerable to this. In 2003–04 nearly 9000 staff of the Department for Work and Pensions had their identities stolen. These were subsequently used in an attempt to defraud the tax credits system in autumn 2005 resulting in £2.7m lost to fraudulent claims. The concern is that this type of successful attack is not unusual. Vulnerability and perceived vulnerability will play a big part in citizens accepting increased technological dependency.

"Despite the difficulties of a fast moving and hostile world, underpinning IT systems must be secure *and* convenient for those intended to use them."—Convenience is a key issue. Currently some of the access procedures that are in place are laborious, complicated and require high levels of literacy and technological confidence. Once connected services tend to be text-based and navigation is often very complicated and lacks intuitive feel.

"Technologies have emerged into widespread use—for instance the mobile phone and other mobile technologies—which government services have yet properly to exploit."—This approach places a cost on the recipient as mobile phone use is never free to those receiving calls and text messages. Some might find the cost inhibiting other might not. Should citizens have to incur costs in engaging with government especially when such engagement is compulsory?

"Over the next decade, the principal preferred channels for the delivery of information and transactional services will be the telephone, internet and mobile channels."—The use of telephone answering menu systems to direct callers to call centre operators is one of society's current irritations. The stress and frustration can be huge of having to drill down through many menu levels to discover that you have to wait for long periods to speak to a real person. This can be intolerable for the vulnerable in need of immediate support from a government agency.

> Sometimes the benefit to society of dealing with government online is not clear. Customer Group Directors and public service providers should also promote responsible channel choice by telling people how much use of more efficient channels saves and what that saving could achieve in terms of reinvestment elsewhere in the public services.

—There is an implication in this approach that technological solutions by their very nature are best. Recent failures of major government systems point to the contrary. For example, the NHS system for booking patients onto waiting lists by GPs is suffering widespread non-take-up by GP practices across the country.

> Government will create an holistic approach to identity management, based on a suite of identity management solutions that enable the public and private sectors to manage risk and provide cost-effective services trusted by customers and stakeholders. These will rationalise electronic gateways and citizen and business record numbers. They will converge towards biometric identity cards and the National Identity Register. This

approach will also consider the practical and legal issues of making wider use of the national insurance number to index citizen records as a transition path towards an identity card.

—There remain huge differences of opinion within society as to the imposition of identity cards regardless of guise or format. (This subject has been covered in previous ETHIcol editions.) Identity validation is important for accessing some government services but it is not ubiquitous. There needs to be clear justification of why and how identification is sought. A relative approach rather than a universal approach would seem to support social justice more comfortably.

There remain many people in society who are socially excluded. These most vulnerable people are likely to have poor literacy, limited communication skills, poor computer literacy, an aversion to technology, a lack of access to technology. The reactive imposition of information technology in the belief that it will automatically improve the help and support of such people must be challenged. An enlightened approach which identifies, develops and implements technological solutions which are fit for purpose and truly promote social justice must be demanded. Overall "Transformational Government Enabled by Technology" appears to move in the right direction but it remains to be seen how such a strategy will be interpreted and implemented; the current track record does not bode well for the future.

5.9 Waking Up to a Surveillance Society [2006]*

This section was first published as: Rogerson, S. (2006) ETHIcol—Waking up to a Surveillance Society. IMIS Journal, Vol 16 No 6, pp. 31–32. Copyright © Simon Rogerson.

The Information Commissioner's Office published in November 2006 "A Surveillance Society'—a detailed report which has been specially commissioned for the conference. It looks at surveillance in 2006 and projects forward ten years to 2016. It describes a surveillance society as one where technology is extensively and routinely used to track and record our activities and movements. This includes systematic tracking and recording of travel and use of public services, automated use of CCTV, analysis of buying habits and financial transactions, and the work-place monitoring of telephone calls, email and internet use. This can often be in ways which are invisible or not obvious to ordinary individuals as they are watched and monitored, and the report shows how pervasive surveillance looks set to accelerate in the years to come.

The Information Commissioner, Richard Thomas said,

> Two years ago I warned that we were in danger of sleepwalking into a surveillance society. Today I fear that we are in fact waking up to a surveillance society that is already all around us. Surveillance activities can be well-intentioned and bring benefits. They may be necessary or desirable—for example to fight terrorism and serious crime, to improve entitlement and access to public and private services, and to improve healthcare. But unseen, uncontrolled or excessive surveillance can foster a climate of suspicion and undermine trust. As ever-more information is collected, shared and used, it intrudes into our private space and leads to decisions which directly influence people's lives. Mistakes can also easily be made with serious consequences—false matches and other cases of mistaken identity, inaccurate facts or inferences, suspicions taken as reality, and breaches of security. I am keen to start a debate about where the lines should be drawn. What is acceptable and what is not?

The report provides a balanced view of the monitored world in which we live. It argues that much of the surveillance in place is a result of well-meaning intentions which become operationally skewed and suffer from function creep. But the report includes concerns about implicit lack of trust in society stating that,

> Most profoundly, all of today's surveillance processes and practices bespeak a world where we know we're not really trusted. Surveillance fosters suspicion. The employer who installs keystroke monitors at workstations, or GPS devices in service vehicles is saying that they do not trust their employees. The welfare benefits administrator who seeks evidence of double-dipping or solicits tip-offs on a possible "spouse-in-the-house" is saying they do not trust their clients. And when parents start to use webcams and GPS systems to check on their teenagers' activities, they are saying they don't trust them either. Some of this, you object, may seem like simple prudence. But how far can this go? Social relationships depend on trust and permitting ourselves to undermine it in this way seems like slow social suicide.

According to the report, "Everyday encounters with surveillance include:

- video cameras which watch us everywhere we go—in buildings, shopping streets, roads, and residential areas. Automatic systems can now recognise number plates (and increasingly faces).
- electronic tags which make sure those on probation do not break their release conditions, and people arrested by police have samples of their DNA taken and kept whether they are guilty or not. "Criminal tendencies" are identified earlier and earlier in life.
- we are constantly asked to prove our identity, for benefits, healthcare, and so on. The UK government now plans to introduce a new system of biometric ID cards, including "biometrics" (fingerprints and iris scans) linked to a massive database of personal information.
- when we travel abroad, who we are, where we go and what we carry with us is checked and monitored and the details stored. Our passports are changing: computer chips carry information, and like ID cards, there are proposals for biometric passports.
- many schools use smart cards and even biometrics to monitor where children are, what they eat or the books they borrow from the library.
- our spending habits are analysed by software, and the data sold to all kinds of businesses. When we call service centres or apply for loans, insurance or mortgages, how quickly we are served and what we are offered depends on what we spend, where we live and who we are.
- our telephones, e-mails and internet use can be tapped and screened for key words and phrases by British and American intelligence services.
- our work is more and more closely monitored for performance and productivity, and even our attitudes and lifestyle outside work are increasingly scrutinised by the organisations that employ us."

A balance must be struck between the legitimate need to undertake surveillance and the desire of us all to live our lives unhindered and unrecorded. It would seem that currently there is an imbalance for, as the report explains, we are all subjected to social sorting in order to define target markets and suspicious populations. Personal data is collected for one purpose and migrated to others without due diligence and consequential consideration of humans. Data endlessly flows across computer networks without little public knowledge of routes and destinations of sensitive data. The report explains that whilst privacy is a key issue regarding surveillance it is not the only issue. The "surveillance society poses ethical and human rights dilemmas that transcend the realm of privacy. Ordinary subjects of surveillance, however knowledgeable, should not be merely expected to have to protect themselves." There are issues of social exclusion and discrimination for "surveillance varies in intensity both geographically and in relation to social class, ethnicity and gender." "Individuals are seriously at a disadvantage in controlling the effects of surveillance" because of colossal power differentials. "Individuals and groups find it difficult to discover what happens to their personal information, who handles it, when and for what purpose."

Professional system developers have a public duty to consider fully the wider implications of surveillance related systems and to ensure that human consequences are understood and that these mitigate the desire to exploit the advances and convergence of technologies in order to achieve economies of scale. Interoperability of surveillance related systems must be justified each time rather than accepted as a de facto standard.

The notion of impact analysis has been discussed previously in ETHIcol. The report argues for the adoption of Surveillance Impact Assessment which promotes the consideration of privacy protection and surveillance limitation from both the individual and society perspective. This impact assessment should be used by professional system developers as part of the methodological approach. Three questions need to be answered:

- "Does the [proposed system] cause unwarranted physical or psychological harm?
- Does the [proposed system] cross a personal boundary without permission (whether involving coercion or deception or a body, relational, or spatial border)?
- Does the [proposed system] violate assumptions that are made about how personal information will be treated, such as no secret recordings?"

By addressing these three questions and ensuring surveillance systems are confined within the boundary of societal acceptability maybe we will live in a free society rather than a surveillance society.

5.10 The Ethics of E-Inclusion [2007]*

This section was first published as: Rogerson, S. (2007) ETHIcol—Ethics of e-inclusion. IMIS Journal, Vol 17 No 6, pp. 31–32. Copyright © Simon Rogerson.

A recent workshop organised by the European Commission examined the ethical aspects of inclusion in the Information Society. The introduction explained that,

> Information and communication technologies (ICT) are advancing rapidly and are changing society profoundly. Such technologies can genuinely empower citizens to play a full role in society, notably those that are at risk of exclusion. Inclusion in the Information Society, or e-inclusion, aims at using ICT to improve social and economic inclusion. E-inclusion is also concerned with inclusive ICT, i.e. ICT that is accessible, available and affordable for all. However, as the knowledge-based economy develops, the increasing use of leading-edge technologies in all areas of life can also present major challenges for some people and introduce new threats to sustained growth and social inclusion. There is a risk that, despite their many benefits, new technologies could set people apart, create new barriers, and increase exclusion.

These ethical issues were summarised as,

> dealing with sensitive problems such as informed consent, right to privacy and protection of personal data, respect for dignity and integrity of the person, non-invasion of the private sphere, equity, intrusiveness (of information technologies), risk and responsibilities for critical technologies . . . [as well as] the use of ICT for social and cultural integration of migrants in their new communities in Europe, ICT for the social inclusion of marginalized young people, and the use of e-government services.

It must be recognised that ethical issues related to ICT are much more than just privacy (albeit this is a very important topic) and that ICT advances will inevitably raise new ethical issues yet to be identified. ICT and human values must be integrated in such a way that ICT advances and protects human values, rather than doing damage to them. This includes both the formulation and justification of policies for the ethical use of ICT and carefully considered, transparent and justified actions leading to ICT outcomes.

This is a daunting and increasing agenda which faces politicians, organisational management as well as ICT professionals. In this age of pervasive ICT, it is not sufficient for ICT professionals to leave the tackling of such sensitive issues to others.

ICT systems are about satisfying a particular requirement or need so that people can realise some economic and/or social objective. In such situations ICT professionals must guard against the design principles where users must adapt to ICT rather than ICT being moulded to users. This is paramount if e-inclusion is to be realised. Ethical design principles must focus on understanding the potential impact on people whose

- behaviour/work process will be affected by the development or delivery of ICT systems
- circumstance/job will be affected by the development or delivery of ICT systems
- experiences will be affected by the development or delivery of ICT systems

The ethical dimension of the ICT has two distinct elements; process and product.

Process concerns the activities of ICT professionals when undertaking research, development and service/product delivery. The ethical focus is professional conduct. It is this focus which is typically addressed by professional bodies in their codes such as the Code of Ethics of IMIS. The aim is for professionals to be virtuous. In other words, a professional knows that an action is the right thing to do in the circumstances and does it for the right motive. Cutting profit so that more development time can be spent on making systems more accessible to those with limited ability, such as dexterity, seems like a virtuous action if it helps to overcome social exclusion.

Product concerns the outcome of professional ICT endeavour. One of the issues of ICT is to avoid systems being used for inappropriate secondary reasons, for example, a security system which has been implemented to reduce the risk of property theft being used to additionally monitor employee movement. Another issue is the thirst of the ICT industry to add more and more facilities in future system releases. Both issues are illustrations of unwarranted function creep. The emphasis should be on accessibility and transparency of systems so people can use ICT systems more easily and can understand, where necessary, how systems work internally. One final issue regarding product is to do with the increasing use of non-human agents based on complex ICT systems. Such agents might interact with humans, for example, those used on the internet to enable e-trading, or they might interact with each other, for example agents which monitor the environment and "order" other agents to take remedial action if necessary.

The ethics focus of the product element is technological integrity. This can be addressed by embedding ethics within ICT products themselves. This might be as simple as building in "opt-in" facilities in e-trading whereby a person must ask to be informed of future sales promotions rather than having to request explicitly not to receive such information by default. They might be more complex, for example, whereby a non-human agent is programmed with defined ethical principles so that it will only instigate actions which are deemed to be societally acceptable.

People deserve ICT systems which will help them fulfil their potential and their goals. Systems which fall short of this by design or by accident will have done so because ICT professionals have failed to take the ethical dimension of their work seriously enough. In today's society we are quick to focus on rights and justice because it is on these which our laws, such as data protection and computer misuse, are based. Whilst they are important there can be a tendency for this to turn into mindless and convenient legal compliance. This will not help e-inclusion whatsoever. Perhaps it is time that ICT professionals adopt care and empathy instead as they grapple with the challenges of ICT process and product.

5.11 Data Profiling in the European Information Society [2007]*

This section was first published as: Rogerson, S. & Cunningham, F (2007) Data Profiling and Personal Data Matching in the European Information Society. Ethically Speaking, Vol 8, July pp. 45–47, European Commission. Copyright © Simon Rogerson & Frank Cunningham.

5.11.1 An Every-day Occurrence

We are all familiar with the irritating question "do you have loyalty card?" when we reach the check-out of our local supermarket. Nonetheless, many of us willingly sign up to an apparently innocent-looking way of gathering "supermarket points" which can be the basis of a powerful form of surveillance and "customer tracking." There are concerns with so much personal information being in the hands of a few private organisations.

Every time we make a telephone call, send an email, use a search engine, use our credit card, withdraw cash from an automatic teller, or even when surfing the web we leave traces of our preferences and lifestyle. Normally when data records are directly associated with a specific name the subject is protected by European and national data protection legislation (but such legislation has a growing list of exceptions and other legislation at times takes preference over this. For example, the European Directive on the retention of data generated or processed in connection with the provision of publicly available electronic communications services or of public communications networks and amending Directive 2002/58/EC). There are an increasing number of ways in which data about us is gathered without our knowledge and consent (e.g., video surveillance in shops and public spaces, the use of RFIDs in supermarkets), which will open new ways of building up a profile of an individual through, for example, places visited, items purchased and carried, and people met.

5.11.2 An Issue Worthy of Detailed Consideration

Digital convergence coupled with this ever-increasing tracking and collection of data about individual European citizens (through, for example, online activities, ambient technologies, RFID and biometrics) have significantly increased the ability to undertake automatic data matching related both to individuals and groups. Whilst such activity might be beneficial and legitimate it can also give rise to some serious concerns about, detrimental effects on privacy, due process of law and presumption of innocence. It can challenge some fundamental human rights (As defined by the Convention for the Protection of Human Rights and Fundamental Freedoms as amended by Protocol No. 11 on 1 November 1998). Actions which appear legitimate under law might not necessarily be legitimate from an ethical standpoint. Once interconnectivity of personal data is established and data matching implemented it becomes very difficult to reverse in the event of a negative impact on citizens or society being identified. It follows that if the European Information Society is to "contribute strongly to improvements in the quality of life" and "reinforce social, economic and territorial cohesion," (COM, 2005) then ICT-enabled personal data matching must be subjected to stringent, *ethical*, social and legal audit.

5.11.3 Data Profiling and Personal Data Matching

Within the context of the European Information Society, data profiling is the process of automatically analysing data to validate it against expected data formats and values and leads to automatic action based on the profiling outcome. It is often incorporated in quality assurance procedures for ICT applications and infrastructure such as ensuring the integrity of data communication networks. Data profiling is also commonly used to match personal data (as defined by the European Directive 95/46/EC on Data Privacy, 24 October 1995) against each other or against a predefined profile. Such personal data is often from a variety of sources and each collected for a different purpose. Thus, data profiling is the key process which underpins and enables *data matching*. Data matching is widespread, providing or prohibiting access, enabling selection, causing inclusion and exclusion. It is used in areas such as marketing (e.g., customer loyalty programmes), employment, financial sector (e.g. anti-money laundering profiling and fraud prevention), forensics, education (e.g. application vetting and e-learning), social services (e.g. benefit entitlement and tax liability), and government (e.g. democracy and voting, and rule of law).

5.11.4 Ethical Concerns about Data Matching

Data matching is used to identify individuals and groups. Using data profiling techniques categories of people are targeted and individuals within those categories identified. The personal data that could be potentially used is huge. For example, it is estimated that in the UK around 115 million electronic files, all with the potential to be accessed for data matching, are held on individual citizens. Across Europe the personal data resource must be enormous as it is likely that a similar number of files exists within each of the 27 member states in addition to those held by central European agencies. The types of data collected are wide ranging. This in part is due to the growth in online activities where personal data is supplied by online users when applying for access or concluding transactions as well as data collected through the planting of cookies. It is in part due to growth in tracking offline activity through scanner data, customer loyalty cards and transaction data of credit/debit cards.

Data matching is initiated not by the suspicion concerning an individual but because the profile of a particular group is of interest. This leads, for example, to three issues.

- *Privacy*—Data matching is likely to involve matching personal records compiled for unrelated purposes which may impact upon personal privacy.

- *Due process of the law*—Once a matching process has been undertaken it will result in a number of positive hits. All those identified are in jeopardy of being found guilty of a wrong-doing. It is unlikely that these individuals are given any notice of their situation, since doing so might affect the investigation, or an opportunity to contest the results of the match at an early stage. For these reasons their right to due process of law is curtailed.
- *Presumption of innocence*—The presumption of innocence is intended to protect people against having to prove that they are free from guilt whenever they are investigated. Data matching can reverse this to a presumption of guilt. This is because the technology of data matching is so plausible and, for example, the detection of fraud is much applauded. These powerful influences will weigh heavy in favour of the notion that those identified must be guilty.

The "permanent" data shadow of citizens left by such files of personal data is concerning and raises many ethical issues (including the three just discussed). Clarke (1994) categorises these as general problems with personal data matching, dangers to the citizen of personal data matching and dangers to society of personal data matching.

5.11.5 A Call for Action

In a technologically-dependent world it is inevitable that every citizen will be subjected to a myriad of personal data collection. There must be proactive socially-responsible action if the curtailment of movement and restriction in unhindered existence are to be avoided. The overt and covert tracking of citizens must be open to scrutiny by society. The linking of unrelated personal data files needs to be properly justified, the reasons for particular data matching need to be explained in detail, there needs to be procedures to severe personal data file links and halt data matching exercises and transparency of data matching should be the norm. Only then will data matching be acceptable to the citizens of the European Information Society.

5.12 Ethics and ICT Governance [2008]*

*This section was first published as: Rogerson, S. (2008) ETHIcol—Ethics and ICT Governance. IMIS Journal, Vol 18 No 3, pp. 27–28. Copyright © Simon Rogerson.

Corporate governance is a widely accepted approach to ensuring that organisations are well coordinated in implementing their corporate visions. Views as to the exact nature of corporate governance vary. For example, the European Commission suggest that it is the "rules, processes and behaviour that affect the way in which powers are exercised . . . particularly as regards openness, participation, accountability, effectiveness and coherence," whereas the Australian Stock Exchange suggest "Good corporate governance structures encourage companies to create value . . . and provide accountability and control systems commensurate with the risks involved."

Following this movement ICT governance is starting to be adopted. Whilst corporate governance is concerned with vertical integration or coordination, particularly at the strategic board level, ICT governance is concerned with horizontal integration or coordination across departments. It is an attempt to ensure joint decision making between IT/IS and line departments and thus countering the silo mentality that pervades organisational ICT.

There does however seem to be shortcomings to this approach. The focus still remains on the drive to be effective and efficient (E2) in the delivery of ICT. This demands engagement with stakeholders but is limited to those within the organisation with direct financial or operational interest. The traditional E2 approach has not a good track record of delivering successful systems in the past so why should ICT governance based on E2 be any different?

What is the missing ingredient? It is ethics. The combination of ethics, efficiency and effectiveness (E3) is appealing. This is because ethics is about understanding right from wrong, efficiency is doing it right, and effectiveness is doing the right thing. Ethics sets the socially acceptable framework within which to operate and moves the perspective beyond only those with a financial or operational interest.

Indeed, E3 provides a new perspective on quality. Not addressing adequately all three elements has a detrimental effect on quality. A quality value derived by some form of multiplication of measurement of the three elements, ethics, efficiency and effectiveness, provides a new way of judging quality. In doing this a proposed system deemed to be unethical will mean it is also deemed to be of poor (and unacceptable) quality and therefore needs to be modified. ICT governance based on E3 and quality based on E3 means that the strategic leadership of ICT is directly linked to the operational implementation of ICT. The goal of delivering fit-for-purpose ICT is more likely to happen by design using the E3 principle.

Consider these two examples of delivered ICT. The first is a robot called ifbot which has been installed in nursing homes to act as a companion for the elderly residents. The robot has the look of an extra from Star Wars with red glowing eyes staring from behind a astronaut's visor. It has a conversational language of a 5 year old. The verdict of this robot is not good. "The residents liked ifbot for about a month before they lost interest. Stuffed animals are more popular" said Yasuko Sawada, director of a Japanese nursing home. "Most (elderly) people are not interested in robots. They see robots as overly complicated and unpractical. They want to be able to get around their house, take a bath, get to the toilet and that's about it," said Ruth Campbell, a geriatric social worker at the University of Tokyo. This piece of ICT maybe efficient in the way it operates, it may be effective in the way it technically delivers synthesised conversations but it is unethical because it fails to take into account the needs of the elderly and treats the elderly in a condescending fashion. It is unfit-for-purpose ICT which would score a negative E3 value.

The second example is an IBM mouse which compensates for the manual tremors that can plague the elderly and those with some forms of physical disability. The mouse treats the hand tremors as noise, and uses algorithms based on image-stabilisation systems used in digital cameras. This is a simple device which has been cleverly thought out and addresses a specific need to improve accessibility for a large group of society. It is impressive how it uses techniques developed for one application and adapted for another. It is fit-for-purpose ICT which would score a high positive E3 value.

Professional ICT development must address both the process of developing an ICT product and nature of the ICT product itself. There are clear relationships between these two perspectives. The manner in which development is undertaken colours the resulting ICT product and similarly the specification of the required ICT product dictates the way product will be developed. Ethics must drive the quest for efficiency and effectiveness. Therefore, ICT governance must address both the process and product dimensions of ICT development. It must promote a sense of obligation in professional developers and ensuring that ICT products are fit-for-purpose.

5.13 The Surveillance Society [2008]*

This section was first published as: Rogerson, S. (2008) ETHIcol—The Surveillance Society. IMIS Journal, Vol 18 No 4, pp. 26–27. Copyright © Simon Rogerson.

On the 20 May 2008 the House of Commons Home Affairs Committee published an important report titled *A Surveillance Society?* which was a comprehensive review of the state of surveillance in the UK and its associated risks and benefits. There is much to be learnt from the report by those who develop and supply ICT components which enable surveillance. It is the advances in ICT which have led to significant increases in actual and potential surveillance activities of individuals in the UK.

A number of key design tenets are included in the report which are based on the principle of data minimisation. It is this principle which helps to keep surveillance in check which is vital as "Loss of privacy through excessive surveillance erodes trust between the individual and the Government [or the private sector] and can change the nature of the relationship between citizen and state [or private organisations]."

It is clear that surveillance in all its many forms casts a very long and permanent data shadow which provides a very detailed view of our lives. Government and its agencies, banks, building societies, credit reference agencies and retailers all add to and use our data shadows. The report points out that "The commercial sector has driven a great many of the developments in this area, recognising the competitive advantage that information about customers can bring when used to target marketing and design personalised services." However, there are significant risks and, "Mistakes in or misuse of databases can cause substantial practical harm to individuals—particularly those who have little awareness of or control over how their information is used."

It seems that we do not fully understand the ramifications of use and dependency on ICT and its long-term effects. This is illustrated by paragraph 9 of the report which states, "Privacy plays an important role in the social contract between citizen and state: to enjoy a private life is to act on the assumption that the state trusts the citizen to behave in a law-abiding and responsible way. Engaging in more surveillance undermines this assumption and erodes trust between citizen and state. In turn such an erosion of trust—with the citizen living under the assumption that he or she is not trusted by the state to behave within the law—may lead to a change in the reaction of the citizen and in his or her behaviour in interactions with other citizens and the Government."

Fit-for-purpose and socially sensitive ICT which "facilitate the collection, storage and use of information about individuals and their activities have clear benefits for the individual as a consumer and a user of public services. If collected accurately and used properly databases of personal information can support both "de-personalised," impartial decision-making processes and the delivery of "personalised" services tailored to the needs of the individual." (paragraph 123) However, there are risks which need to be carefully addressed which include the erosion of privacy and individual liberty, the excessive amount of personal information collected, the excessive length of time such information is kept, and the lack of awareness or ability of an individual to check and control information about them. Paragraph 76 of the report points out that "A strong common theme is emerging in both the private and public sector: a move towards more personalised services which require the service provider to collect information from individuals in order for the service to be effective. Whilst the outcome may be more personalised, however, the trend in terms of input is a standardisation of the information requested with a tendency to collect information which may identify an individual even where this is not needed in order to provide or improve services."

Three practical tenets provide a framework within which to develop surveillance information systems.

- "The principle of restricting the amount of information collected to that which is needed to provide a service should guide the design of any system which involves the collection and storage of personal information." (paragraph 163)

- "Information should be held only as long as is necessary to fulfil the purpose for which it was collected. If information is to be retained for secondary purposes rather than service delivery it should normally be anonymised and retained only for a previously specified period." (paragraph 164)
- "In keeping with a principle of data minimisation, more rigorous risk analysis of systems already in place must be carried out before new techniques for collecting information are deployed or new databases planned. The decision to create a major new database, share information on databases, or implement proposals for increased surveillance should be based on a proven need." (paragraph 190)

Without ICT the surveillance society would not exist. Every ICT professional must review his or her role in sustaining a society where surveillance is sufficient to be beneficial for each of us and is not excessive so it undermines trust, self-esteem and freedom.

5.14 Identity and DNA [2008]*

This section was first published as: Rogerson, S. (2008) ETHIcol—Identity and DNA. IMIS Journal, Vol 18 No 5, pp. 28–29. Copyright © Simon Rogerson.

Scientific advances continue to amaze. No more so that the work in DNA and genetics. The question must always be asked of how a scientific advance will help humankind and the world at large. Recently the increasing use of DNA and genetic information to establish identity has come under public scrutiny. The Human Genetics Commission has undertaken a study of citizens" attitude towards the UK National DNA Database. This database is the largest of its kind in the world holding data of over one million people including 100,000 children. Genetic material is taken from all people arrested whether they are then charged and convicted. This data remains of the database forever. Baroness Joan Walmsley has concerns about this database. She points out that,

> The Government is collecting more and more information about us, but seems utterly incapable of keeping it safe. . . . One of the fundamental tenets of British justice is innocent until proven guilty. By refusing to destroy samples from those who are never charged with a crime or who are later acquitted completely blurs that principle.

Such concerns are shared by others. The HGC study found the majority of participants concluded that:

- the retention period for profiles on the database should be proportionate to the seriousness of the crime for which the person to whom they relate was convicted
- the ethnic group of a person from whom a DNA sample was taken should not be recorded as this could contribute to discrimination
- samples should be destroyed and profiles removed from the NDNAD when a suspect is not proceeded against or an accused person is not convicted at the conclusion of criminal proceedings

Baroness Walmsley continued "We also call for the removal of all DNA samples of children under the age of 16, except those who have been convicted of a violent or sexual offence. Placing children on the NDNAD may actually increase their propensity to commit further crime, making the retention of their samples entirely counter-productive." This sentiment is echoed in the HGC report.

The collection of ethic profiling data has raised much concern. Christian Today (6 September 2008) reported on research by the human rights campaign group Black Mental Health UK which has shown public opinion to be very strongly against the practice of adding the DNA and ethnicity data to the National DNA database of mental health patients who come into contact with the police but have not committed a crime. Health experts have warned this is criminalising one of society's most vulnerable groups. The Rev Paul Grey was reported as saying,

> Why are people on the database if they haven't committed any crime? People needing mental health care are vulnerable and when people are vulnerable it is important to protect them. Part of the freedoms and civil liberties we have in this country is that we are able to keep what belong to us, and there is nothing more personal than our DNA.

Questions of access and anonymity surround DNA data as well. The Contra Costa Times (28 Aug 2008) reported privacy concerns over a database of so-called anonymised DNA profiles of 60,000 volunteering patients had led to the National Institutes of Health in the US blocking public access to this database.

> A new type of DNA analysis could confirm the identity of an individual in a pool of similarly masked data if that person's genetic profile was already known. Such a confirmation could reveal the patient's participation in a study about a specific medical condition, experts said, and deny the patient their presumed confidentiality.

It is clear that advances in genetic science are leaving policy and legislation in its wake. Fred Bieber, a medical geneticist at Harvard Medical School was reported as saying, "The lesson is that with enough genetic information, it's becoming easier to identify individuals even though their identities are presumed to be made anonymous."

These issues surrounding DNA databases are about trust. It is about the relationship of citizens with government and its agencies. How much confidence do citizens have in their government to collect, store, and use DNA and genetic information in a way which is acceptable to us? Such trust is not simply based upon an intellectual assessment but an emotionally experienced sense of security and assurance. Both are equally important. Our measure of trust is based upon what we think of the other party's benevolence, honesty, sincerity, openness, and caring. To date those responsible for the policy and operation regarding databases such as the National DNA database do not seem to be faring well.

5.15 Responsibilities in Product Creation [2011]*

This section was first published as: Rogerson, S. & Miller, K. (2011) ETHIcol—Responsibilities in Product Creation. IMIS Journal, Vol 21 No 1, pp. 36–37. Copyright © Simon Rogerson & Keith Miller.

As ICT becomes more advanced and the application reach increases, the complexity of the ICT products created becomes more and more complex. Every issue of the IMIS journal will have at least one article about problems with these complex products, which clearly suggests that this is an issue that should concern us all. This issue of complexity has become the focus of a group of international experts who have come together to develop a set of guidelines to help ICT professionals understand their moral responsibilities regarding the ICT Products. The guidelines are titled *Moral Responsibility for Computing Artifacts* and are discussed in this edition of ETHIcol.

The term *computing artefacts* is synonymous with *ICT Products* and is defined as anything which includes an executing computer program. The list is therefore large, from software applications running on a general purpose computer to programs burned into hardware and embedded in mechanical devices, robots, phones, webbots and toys. All software falls within this definition regardless of whether its origin is commercial, free, open source, recreational, an academic exercise or a research tool.

People are answerable for their actions or behaviour when they produce or use computing artefacts. They have obligations to adhere to reasonable standards of behaviour and to respect others who could be affected by their actions. It is clear that ICT professionals must recognise this and act in an acceptable manner when producing any computing artefact. This can be a difficult challenge for many reasons, and the guidelines are an attempt to offer some support. The guidelines centre around five rules.

Rule 1 includes, "The people who design, develop, or deploy a computing artifact are morally responsible for that artifact, and for the foreseeable effects of that artifact." Whilst **Rule 2** includes, "The responsibility of an individual is not reduced simply because more people become involved in designing, developing, deploying or using the artifact."

These first rules recognise that people consciously and intentionally produce computer artefacts and therefore must be accountable. The abdication of responsibility through the claim that an individual's responsibility is insignificant when an artefact is produced by many people is explicitly rejected. It is important that all ICT professionals and all levels understand this.

Whilst it is fair to acknowledge that people cannot reasonably be expected to foresee all the effects of the computer artefact, for all time, it is equally fair to expect that people make every effort to predict the uses, misuses, and effects of the deployment; and to monitor these after deployment. Wilful ignorance, or cursory thought, is not sufficient to meet one's moral obligation. If people produce a computer artefact in such a way that it is not possible to reasonably predict its future behaviours, then they are particularly responsible for the unpredictable, and potentially harmful, results.

Some machines are designed to adapt over time, to learn without human supervision, or to self-modify their code. In these cases their producers have a heavier burden of responsibility. People who recognise their responsibilities are likely to make their machines simpler and more predictable in order to make them safer and more reliable.

Rule 3 states, "People who knowingly use a particular computing artifact are morally responsible for that use." The moral responsibility of a user includes an obligation to learn enough about the computing artefact's effect to make an informed judgment about its use for a particular application. At first glance this might appear to be a rule for ICT users rather than ICT producers but that is incorrect. For example, ICT producers use development computer artefacts to produce application ICT artefacts. If the limitations of such development tools are not understood, then the resulting application computer artefact could be flawed.

Rule 4 states, "People who knowingly design, develop, deploy, or use a computing artifact can do so responsibly only when they make a reasonable effort to take into account the sociotechnical systems in which the artifact is embedded." This is an attempt to remind us that ICT does not exist in isolation; it exists in an environmental context (called here the sociotechnical system). People who ignore the context are morally irresponsible. This rule is intended to be a progressively heavy burden. Thus, the burden is heavier for those with more expertise of and more influence over the effects of the computer artefact and the environmental context.

Finally, **Rule 5** includes, "People who design, develop, deploy, promote, or evaluate a computing artifact should not explicitly or implicitly deceive users about the artifact or its foreseeable effects." Morally responsible use of computing artefacts requires reliable information about such artefacts. People who produce or promote a computing artefact should provide honest, reliable, and understandable information about the artefact, its effects, and possible uses and misuses. People are morally responsible for their choice of whom to inform, and for what they disclose or do not disclose.

Regardless of how sophisticated the computer artefact these five rules still apply and provide clear guidance to ICT professionals at all levels as to their moral responsibility in their professional practice as they strive to satisfy the needs and address the problems of society and its citizens. To ignore this guidance is foolhardy and unprofessional, and is likely to result in a more dangerous and more problematic Information Society.

5.16 Ethics of ICT [2012]*

This section was first published as: Rogerson, S. (2012) ETHIcol—Ethics of ICT. IMIS Journal, Vol 22 No 3, pp. 36–37. Copyright © Simon Rogerson.

In February of this year (2012) the European Group of Ethics in Science and Technologies (EGE) published *Opinion No. 26, Ethics of Information and Communication Technologies.* This was at the request of President Barroso who suggested that the Opinion could offer a reference point in the promotion of the Digital Agenda for Europe. This request in itself is indicative how the ethical dimension of ICT development and application is recognised, by society in general and politicians in particular, as being paramount.

The Digital Agenda implies the need for an ethical approach when the key policies of the five year plan are considered. These policies are summarised as:

■ Create a single market to remove trade and licensing barriers and simplify copyright.
■ Improve ICT standard setting and operability.
■ Increase access to fast internet and wireless networks.
■ Raise levels of digital literacy throughout society.
■ Apply ICT to major societal challenges such as climate change and the ageing population.

Opinion 26 attempts to tease out the ethical approach. As such it is of a general nature covering a range of issues including access, identity, e-commerce, privacy and data protection. The issues discussed have global relevance. The Opinion is a good reference for all ICT professionals and should become mandatory reading material.

A set of fundamental principles supporting peace and well-being are reemphasised which are significant for ICT. These principles are:

■ human dignity
■ respect of freedom; which demands uncensored communication and agency
■ respect for democracy, citizenship and participation; which demands protection against unjustified exclusion and discrimination
■ respect for privacy; which demands protection of personal space from unjustified interventions

- respect for autonomy and informed consent
- justice; which demands equal access to ICT and the fair sharing of the associated benefits
- solidarity; which secures social inclusion in ICT access providing choice for all

The Opinion culminates in a series of recommendations of which some are discussed here.

There is an emphasis on increasing digital awareness and literacy with a call for educational programmes to enable individuals, especially those with special needs, to develop technical and digital literacy. There should be greater focus on respect, tolerance and sensitivity when communicating digitally. Educational tools for simplifying ICT applications and internet usage are suggested. Furthermore, it is recommended that educational programmes are operated which foster and raise awareness and responsibility regarding the impact of ICT on personal, social and moral identity. There is a strong message to service providers that responsibility should be fostered regarding accountability, identification and traceability for internet identities. In this way society will be more trusting of the online world and those who provide access to it.

There are several recommendations regarding the right to privacy and the protection of data. The Opinion states that, "transparency is a fundamental condition for enabling individuals to exercise control over their own data and to ensure effective protection of personal data." For this reason, it is paramount that individuals are provided with clear information in a simple and transparent manner. Furthermore, there is a call for explicit consent of those whose personal data is being processed. It is recommended that the ability must always exist to withdraw consent without any negative consequences for the data subject. Indeed, "Individuals should be explicitly informed by businesses, State bodies or research bodies that their information may be mined for specific purposes. This will ensure that individuals can make informed choices about the services they access and use. Specific consent should always be sought when databases are correlated."

In support of the right to be forgotten, the Opinion states that, "the right to deletion of personal data should be extended in such a way that any publicly available copies or replications should be deleted." With advances in science and technology a number of fresh issues have arisen regarding sensitive data. It is recommended that categories and conditions of processing should be reviewed. For example, it is explained that the provisions regarding genetic data and biometric data are unclear.

In total, these privacy recommendations place the emphasis on safeguarding individuals rather that supporting ease of processing. This seems more appropriate in a technologically-dependant society. It has serious ramifications of ICT processionals working not only on existing application systems by also new developments.

There is a strong emphasis on the social aspect of ICT within the Opinion. Two issues; digital divide and work life balance are particularly interesting and are reproduced here:

> The EGE recognises that disadvantaged and marginalised groups may require different designs, content and applications to suit their specific requirements. To this end, the EGE recommends that measures centred around direct provision, subsidies and regulation be examined by the EU to ensure that such groups are not excluded from playing a full and active role in the digital society.
>
> The EGE urges the EU to encourage and support organisations to develop explicit policies to ensure the optimal use of ICT while respecting the work life balance. Such policies should aim to foster an organisational culture which does not create an expectation that employees should be "on call" during non-work hours. This should also be considered in the corporate responsibility programmes and labour regulations.

Citizen empowerment is a powerful theme within the Opinion with a demand that the online world is a free world. It states,

> the need for keeping the Net a free and neutral space. This freedom must not contravene the fundamental ethical values of the EU. . . . The internet must remain a communication domain where freedom of expression is protected from censorship. . . . [There is] the need to balance top-down internet governance by governmental agencies with bottom-up participatory approaches by the internet community.

Opinion 26 is provocative and challenging for the ICT professional. Rather than wait for regulation to be imposed which forces the ICT sector to adopt a new social perspective it would be refreshing for all ICT professionals to grasp the initiative and from today start to operate in alignment with all the recommendations within the Opinion.

5.17 Ethics and the IT Profession [2017]*

This is section is based on a transcript, albeit slightly modified, of Ethics and the IT Profession, *an interview of Simon Rogerson ([A] in the transcript) conducted by Dr Don Heath ([Qn] in the transcript)on 28 November 2017 for an online course,* Legal and Ethical Responsibilities of IT Professionals *which is part the Applied Computer Science degree offered by the University of Wisconsin, US. Copyright © Simon Rogerson & Don Heath.*

[Q1] What is the importance of ethics to IT professionals?

[A] The thing we need to remember is that IT professionals plan, develop, and deliver products and services for the world at large. And as such, they have obligations and responsibilities to all who are directly and indirectly affected by these products and services. So therefore, an IT professional's actions must be ethical and the resulting outcomes of these actions must be ethically rigorous. IT simply cannot happen in isolation from society. And the IT industry is littered with examples of questionable practices, such as moral algorithm developments and use without due consultation, public beta testing of safety critical software, and the inherent risks of legacy systems, function creep in advancing IT such as big data and the internet of things.

As society becomes more and more technologically dependent, so, in my view, it becomes more and more technologically vulnerable. The conduct of IT professionals is essential to this paradoxical situation. So ethical IT promotes social societal benefits. Unethical IT leads to societal damage, often incurred by its most vulnerable members. And that's why I think it is an important area for all IT professionals.

[Q2] Can you outline the code and explain the motivation behinds its creation?

[A] In 1996, I think it was, Don Gotterbarn and I met for the first time at PACE'96 conference, which was organised by the University of Westminster in London in the UK. Don was there as a representative of the ACM, and with the help of Keith Miller, he had been instrumental in getting the ACM to agree to develop a tailored code for software engineers. The aim was to provide ethical guidance, which would resonate with practitioners, and I joined Don and Keith in leading the project.

The three of us have practical IT backgrounds and we share common beliefs that ethical guidance and training must have practical worth if the so-called IT profession is to be recognised as a profession both in name and in deed. Don spent a six months sabbatical working on the code at the Centre for Computing and Social Responsibility at De Montfort University in the UK, where, as you'd mentioned earlier, I was the founding director. The project ran for about three years, with input from many people, worldwide.

An agreement with IEEE-CS was reached, which led to the code being adopted by both professional bodies in 1999, quite some time ago. What's interesting is that the "Software Engineering Code of Ethics and Professional Practice" remains current. It's been translated into many languages, as well as being adopted by many professional bodies, and public and commercial organisations across the world. So, it's a very important code, and one which has resonated throughout the world and, as I said, remains current today.

[Q3] How have circumstances changed since the Software Engineering Code of Ethics was created?

[A] I'd like to start with a couple of examples. It's a bit of a history lesson, really. So, as a young Fortran programmer in the 1970s, I was instructed to incorporate a covert time bomb into a design system to be rolled back to a subsidiary company. Now I saw nothing wrong with building these functions. In fact, it was a fascinating technological challenge. Ethics did not occur to me. I was told to do it by the most senior member of staff in the department, and I was just a junior programmer. I had never been made aware of the ethical dimension of IT. I was, indeed, an uneducated technologist. So, if we fast forward to January of this year, when VW pleaded guilty to using defeat software to cheat on mandatory emissions tests, as well as lying and obstructing justice regarding the future of the scheme, it's very interesting—at the centre of that scandal, in my view, is misinformation generated by on-board software. This system was developed and implemented by IT professionals who must have been party to this illegal and unethical act. Professional practice is unprofessional without ethics, yet here is a time span of nearly 50 years, which suggests that the industry has learned very little about the importance of ethics related to IT.

Indeed, BCS, the Chartered Institute of IT, based here in the UK, ran a special edition of ITNOW in the autumn of 2014, I think it was, which focused on ethics of IT. In many ways, it's a litmus test of ethics progress by practitioners and academics working in tandem. Sadly, it is a disappointing read. The lack of ethical consideration in system design and implementation is evident. The calls for action are neither new nor inspiring. There is virtually no evidence and no pragmatic action. The emphasis being very much on top down political rhetoric, and we currently have too much of that across the world. In many ways, this edition illustrates, at best, that we've stood still. But probably, we're moving backwards in the quest for ethically acceptable technological implementations.

Now, even though there are many excellent codes created and disseminated by professional bodies allied to IT, why is it that significant unethical activity within IT remains? I think one reason might be that the current focus of effort and attention is inappropriate. IT development is a global activity involving around 11 million professionals and 7.5 million hobbyists. And only around 1% of that total belongs to a professional body. Now this suggests, on the basis of statistics, that professional bodies allied to IT and their adopted codes have little influence on practical IT.

[Q4] Can you talk about your current research and publications?

[A] Well I'm supposed to be retired. But there we go. I don't think anybody who's involved in this sort of area will ever be retired. I've always been interested in the practical worth and development of IT artefacts, and their application where such artefacts have been used in both planned and unplanned situations.

So, I continue to do that, and I'm currently working on two application areas—transportation and tourism, which I've not previously considered. I've just had a paper accepted for the February edition of the Communications of the ACM, relating to the VW emissions scandal, which I mentioned earlier. And I'm working on a conference paper which will propose a technology taxonomy for the tourist industry, demonstrating recurrent ethical challenges over time. Alongside that, I'm developing a historical record of the ETHICOMP Conference Series as it mirrors the advances in ethical thinking and action related to computing since 1995. And my plan is that this will be available on a dedicated website, and I think there is much for practitioners and scholars to learn from that conferences series and what was done in that conference series from 1995.

[Q5] Are there social or technology trends that you believe will change the way we think about ethics and IT?

[A] In general, if I reflect on computing from when I first got engaged in this in the early '70s, society has become accepting of IT. Such things as social media, internet shopping, smart phones have entered our everyday lives, and this trend will continue as the interface between us and IT, I think becomes easier. So, the traditional interfaces of keyboard, screen, and mouse will be completely replaced as machine vision, voice recognition and generation, and connectivity evolves.

Linked to this is what I consider to be one of the greatest challenges for society, and that is the online world. Technology of the online world is continually evolving, providing new tools and experiences. Currently, individuals can choose how much or how little of the online world they wish to engage with. However, such things as peer group pressure, social norms, mass media, government policy, as well as commercial marketing all influence what is available and the choices made.

It is likely that as technology evolves, eventually, online engagement will be mandatory to us all. The more sophisticated technology becomes, then the experience is extended, and intimacy increases. And these experiences change our perception of self, place, action, and belonging. In some senses, I think humans are becoming composite beings. If you acknowledge this, then that leads to the need to think of us not just as data subjects, but as data selves. And it's interesting that the new European legislation on privacy is starting to reflect that movement from data subject to data self. Data—the virtual autonomy, should not be owned by third parties. Now that's controversial.

The physical building of trust relies on visual cues, which are not accessible online. And so, alternative, effective ways need to be found to establish online trust. The use of texting and emoticons, for example, by ordinary people, seems to have substance. For them, the social scaffolding, such as the Declaration of Human Rights needs to be reviewed and revisited in the light of composite human beings. Overall policy, regulation, social norms need to be revisited, modified, and enhanced to account for this technological evolution, which is also an ethical evolution. And I think that the IT professional has a significant role to play in shaping this new world.

[Heath interjects] *That's fascinating, and you know, it's certainly the case that we're embedding our own ethical understanding in some of the software that we produce now, which adds to that convergence.*

That is very true. Yes, there is this whole business about what I call process and product. You know, it is the doing of IT, and then it is the real resulting IT outcome. The challenge is, exactly as you say, how to embed the ethical, social norms into the IT that you've created.

[Q6] What are the ethical responsibilities of IT professionals?

[A] In terms of the IT professional, I think there are probably three key responsibilities, from an ethical perspective. I think the first one is that IT professionals should have the moral courage to do the right thing. And in my book, this is a fundamental principle for professionals, in general. They should adhere to their code of ethics in a reflective, rather than a compliant manner. And currently, there is a lot of pressure for us to comply with things without actually believing in them or doing them properly. And that, I think is a backwards step. So be reflective, rather than compliant.

The second thing is I think IT professionals should be proactive in promoting ethical behaviour and challenge complacency, indifference, and ambivalence regarding ethical IT by those involved in any aspects of researching, developing, implementing, and using IT. So, this means interacting with any of the 18.5 million practitioners I mentioned earlier.

And I think the final important key responsibility is that established IT professionals should mentor early career members in the ways of ethical IT. So, we have some future proofing, in terms of being ethical, as a profession.

[Q7] What rights should be extended to IT professionals and those they interact with?

[A] That is a very interesting question which is rarely considered. Let me try and answer that in several parts, and again, I'll focus on the ethical perspective. So generally, I think that it is the right to be involved in technological change discussion and decision making at all stages. And why I say that is because it is the IT professionals who are technologically savvy, and if they're educated properly, they will have contextual understanding. And quite often, decisions are made by people who really don't understand the technology, and that is a problem. So, I think that's generally a general point.

In terms of the employer, I think there are probably three things that I would say. There's the right to have justifiable ethical concerns investigated and acted upon. There is the right not to be pressurised to act in an unethical manner. And I think there is also the right to receive multidisciplinary professional development. So, we have no longer got uneducated technologists, as I alluded that I was in the 1970s. And if we look at the client, I think there's the right to receive requests which are reasonable, in terms of scope and expectations, or to have unreasonable requests modified. So, it isn't—you know, things go wrong—it's always the fault of the IT professional. And quite often, that may well not be the case. It might be, simply, the fault of miscommunication.

And finally, there is the issue of the user. And I think that's quite simply the right to have technological change viewed with an open mind. How many times do users not want to embrace a new way of doing things? Understandably so. But it's about having an open mind, and I think that then makes the IT professional feel more valued and more willing to then enter into a partnership of equals between those they're engaging with and themselves.

[Q8] How are ethical considerations impacting what it means to be an IT professional?

[A] Well, the first thing we need to remember is that IT is no longer restricted to the back office. Indeed, it is no longer restricted to the organisation, as a whole. It is pervasive. The range and complexity of the surrounding ethical issues of pervasive IT increases almost daily. In order to be efficient, effective, and ethical, the IT professional must understand, for example, different application areas, different cultures, different disciplines. Therefore, training and ongoing professional development must not only be technical, but it must be multidisciplinary. It is not possible simply to be a narrow minded technologist in this new age and consider yourself to be an IT professional.

[Q9] If you were to offer a single piece of advice to those entering the IT profession, what would it be?

[A] For me, that is relatively straightforward. Always remember, according to the "Software Engineering Code of Ethics and Professional Practice," the concern for health, safety, and welfare of the public is primary. So, whenever you are undertaking a piece of IT work, set your perspective as far away from the technology as possible. In this way, your scope of consideration is more likely to include the issues and people deemed to affect or be affected by this work. The chance of identifying any associated ethical issues, then, is greatly increased. And you will be able to act as an informed, ethical, IT professional.

References

Clarke, R. (1994). The digital persona and its application to data surveillance. *The Information Society, 10*(2), 77–92.

COM (2005) 229 Final—i2010—A European Information Society for Growth and Employment.

Matthews, M., Rice, F., & Quan, T. (2021). *Responsible innovation in Canada and beyond: Understanding and improving the social impacts of technology.* Information and Communications Technology Council.

Rogerson, S. (2021). *The evolving landscape of ethical digital technology.* Taylor & Francis Group, ISBN 9781003203032.

Chapter 6

Practical Considerations

Previous chapters have taken specific perspectives. In this chapter the emphasis is on holistic issues. A range of practical challenges are included which will encourage the reader to reflect on how best to practise in a societally acceptable manner. Two recent examples highlight the necessity to be practically vigilant in the implementation of digital technology.

The first example relates to the Y2K problem, which has been previously discussed in Section 4.6. However, the same fundamental systems design flaw recently occurred in a completely different application related to the COVID-19 pandemic. DeBré (2021) reported on how a software error had incorrectly shown an alarmingly high mortality rate in children living in Spain. She explained that Bhopal et al. (2021), writing in the well-respected *Lancet*, had reported the high mortality rate which raised huge concerns internationally. Within days the integrity of the findings was questioned by Dr Pere Soler, who is a paediatrician at Hospital Universitari Vall d'Hebron in Barcelona. This led to a reanalysis of the data to discover there was not a high mortality rate in Spanish children. The *Lancet* republished the corrected findings on 24 March 2021. The reason for the error was that the data analysis software could not record three digit numbers and so, for example, a person dying at the age of 102 was recorded as a child of 2 years of age. This was the exact system flaw which caused the Y2K crisis. Fortunately, in this recent case, the error was spotted very quickly; otherwise, it could have resulted in effort and resources being wrongly focussed during the global pandemic.

The second example concerns Facebook, one of the pillars of the modern digital age, with its 3.5 billion users worldwide as well as 1.7 billion users of Instagram, which it owns. Recently, former Facebook engineer Frances Haugen, who worked as a product manager on the civic integrity team, triggered an international investigation into the internal workings of Facebook. Before leaving the company, she copied a series of internal memos and documents which she then shared with the *Wall Street Journal*. These documents were then published, revealing many societal concerns regarding the workings and policies of Facebook. For example, Murgia and Espinoza (2021) reported that Haugen

> claimed Facebook's algorithms can quickly pull people down into psychological rabbit holes. "In the case of kids, they can follow very neutral interests like healthy eating and be drawn into anorexia content," Haugen said. "That is not illegal, but it is really harmful and kids die as a result of those things." . . . "Whistleblowers are only going to get more and more important for society—technology is accelerating ever faster and

DOI: 10.1201/9781003309079-6

the only people who understand these technologies are people inside the companies," Haugen said.

It remains to be seen what, if any, ethically effective changes will be made to Facebook and Instagram in the light of this whistleblower's revelations.

6.1 Teleworking [1997]*

*This section was first published as: Rogerson, S. & Fairweather, N.B. (1997) ETHIcol—Teleworking. IMIS Journal, Vol 7 No 2, p. 29. Copyright © Simon Rogerson & N Ben Fairweather.

Increasingly work is being conducted in projects, with employers asking greater flexibility of their employees, including frequent changes of workplace. Even with businesses increasingly based in "accessible locations," in so many of sectors of industry, including the computing professions, frequent relocation is the norm. All this means staff are put under great stress, commuting long distances, and uprooting themselves and their families regularly. This stress also has adverse impacts on the business, with loss of productivity and increased time off work sick. The time spent commuting means that staff are liable to be more tired and less willing to work long hours, while using cars and the road network to commute increases pollution and the chance that the country will not meet international obligations to reduce carbon dioxide emissions. At the same time, employers may be particularly anxious to retain key employees (usually women) who have parenting duties that need to take priority over work at times. Increasingly, also, employers are under pressure to not discriminate against disabled people, some of whom cannot easily be accommodated in existing offices and ways of working.

One apparent way out of these binds that may even increase flexibility further is to use teleworking—combining computing and telecommunications technologies to work from an office at home (or in the locality of the home) and away from the "normal" workplace. There are right and wrong ways to organise telework; but if handled properly, industry can gain through reduced costs, society can gain through reduced levels of road traffic and strengthened communities, workers can gain through not having to pay the financial and physical prices of commuting, workers" families can gain by not having to move to different towns frequently, disabled people can gain new access to work, and parents can combine work and childcare in new, more flexible, ways.

A full list of the potential benefits of telework would be enormous. So, what sorts of jobs can be teleworked? Rather than attempting to produce a complete list, it is worth starting with consideration of the sorts of work that cannot practically be converted to telework.

Work that has a high element of physical labour clearly is not suitable for teleworking. Similarly, because telework will be done away from the main base of the employer, work that requires physical use of large and expensive equipment cannot practically be teleworked. Work that already is centred around the computer and telephone is a prime candidate for telework, so information technology specialists such as software programmers and systems engineers might be able to telework. But the possible range of jobs that can be teleworked goes far beyond this and includes professional and management specialists such as financial analysts, accountants, public relations staff, graphic designers, and general managers. Similarly, such support workers as translators, bookkeepers, proofreaders, draughtsmen, researchers, data entry staff, telesales staff and word processor operators; could all telework, given suitable circumstances and arrangements.

There are yet further applications of telework possible with recent technological developments. Professions that are now thought of as being "hands on" can increasingly be conducted "down the line': thus, surgeons have conducted operations by remotely controlling robotic tools in a hospital hundreds of miles away. Equally, banking, once thought to involve large numbers of staff sitting in local branches, is increasingly being conducted by telephone rather than face-to-face contact. With imagination, huge numbers of jobs could be teleworked.

But what, exactly, are the advantages to a business of making use of teleworking? Teleworking gives a chance to substantially reduce the amount of office space needed. A 1994 survey for British Telecom showed that City firms could save £21,000 pa for each senior manager teleworking: other businesses will not benefit quite so much, but will still save thousands of pounds each year for every teleworker. Productivity is, on average, higher among teleworkers, with about half of employers of teleworkers reporting productivity increases, and only one in twenty reporting reductions. Studies also report an average reduction in days off "sick" of two days per year. There is also a possibility of recruiting and retaining trained and talented staff that would not otherwise be available for employment.

There are, however, possible disadvantages for the business of teleworking. Not every worker is suitable for teleworking—some need set working hours in an office away from home to work consistently. Teleworking requires management by results, rather than the "over the shoulder" methods that managers may be used to (but this change in managerial culture may be useful anyway). Equally there are potential problems for teleworkers that responsible employers need to guard against. Prime among these are isolation, and lack of involvement in decisions: many teleworking schemes thus involve the teleworker being at the main office one day a week (or so). Similarly, unless specific measures are taken to avoid the problem, teleworkers may be at a disadvantage when promotion becomes available. Some employers fearful of the loss of control over staff have monitored, for instance, the number of keystrokes per minute that an employee is making, in ways that make thinking time seem like absenteeism: employees often find such arrangements intimidating and intrusive: management by results would render them unnecessary anyway. There are other possible disadvantages. Disabled people worry that the availability of some telework may be used to deny them access to mainstream employment, or as an excuse for not making improvements to premises and transport systems that fairness and participation in society will still require. Equally, some employees find the loss of physical distance between work and home leads to unwelcome intrusion, so responsible employers might need to provide dedicated telephone lines with answerphones for after-hours calls, or let workers stop teleworking if things are not going well.

While there are possible disadvantages, teleworking has many greater potential benefits for employers, the community at large, and employees. The challenge is to take the initiative to implement it widely, and in ways that are fair to all.

6.2 Alternative Views [1997]*

*This section was first published as: Rogerson, S. (1997) ETHIcol—Alternative Views. IMIS Journal, Vol 7 No 5, p. 11. Copyright © Simon Rogerson.

In this issue we look at a range of alternative views concerning the application of information technology in today's society. Whilst some might argue that these views are both extreme and negative they do provide an antidote to the over-inflated claims often found in the press about the benefits of the Information Society and how it will liberate and harmonise the world at large.

In a recent article, Wayne Ellwood (1996) of the New Internationalist argues that computers are changing the world and not always for the better. For example, computers have enabled money speculators to make huge profits through daily foreign exchange trading estimated at

$1.3 billion. This destabilises economies and decimates long-term investment. Electronic war is now a reality where human decision making has been superseded by computer programs as was the case during "Desert Storm." Ellwood argues that this has increased the likelihood of a major catastrophe. Employment patterns have radically altered with the advent of computers. There now exists two groups of workers; those who are well educated, experienced and relatively well paid and work long hours of overtime, and a much larger group of low paid, unskilled and part-time workers who can be considered as being interchangeable. Ellwood points out that there is also a growing underclass of the chronic jobless and socially marginalised. Given the existence of such powerful examples, he advocates that the process of introducing IT into society and the workplace should be democratised so that each citizen can have an input in deciding if, when and how IT should be applied.

Turning to education, Theodore Roszak has serious reservations about using the internet in the classroom. He argues that the search engines of the World Wide Web are distracting with their brash advertisements and tempting free offers. The results of a search usually produce thousands of "hits," many of which are irrelevant. It takes considerable time and skill to hone a search and this task is often beyond the capability of a young pupil. Roszak argues that the internet is a free-for-all with no quality control which a school library would be subject to, and that "the Web is an expensive way to distract attention and clutter the mind" and so is inappropriate as an education resource. Roszak urges us to remember that education predates high technology.

In a thought provoking article, the author Kirkpatrick Sale explains that whilst there are many good things in the Information Society there is also a dark side. He points out that the computer revolution has contributed to widening the gap between the haves and have-nots. In the US 5% of the population own 90% of the wealth and only half the working population have fulfilling jobs. This situation has been exacerbated with the introduction of computers and computer-controlled processes in the workplace. He points out that each year 200,000 tons of poisonous wastes are created by computer chip manufacturers and that there are 25 million on-the-job injuries every year by people who use computers. Sale suggests that we have become oblivious of such detrimental effects and prefer to focus on the materialistic personal rewards that IT offers us.

There are many who would argue that IT has presented great economic opportunities for the poor and developing countries of the world. At the forefront of this economic revolution is the Indian software industry with its centre in Bangalore. In 1995 it had sales of $2.2 billion and this is growing at more than 40% each year. It has flourished due to high quality Indian software engineers, low wages and existing in a different time zone to its American clients. According to Dilip D'Souza this a false dawn. Western companies and the 10,000 software engineers in Bangalore might prosper but half the country remains illiterate and six out of ten people in the cities live on the streets or in slums without water and sanitation.

The eminent physicist, Ursula Franklin has been an outspoken critic of the impact of technology on society. She suggests that economic efficiency and profit dominance have resulted in technology becoming the control agent where individuals have to comply with the technological norms. She argues that policy regarding technology application should be founded upon socially favourable responses to the following questions.

Does the technology:

■ promote social and economic justice?
■ restore reciprocity or consolidate power in the hands of a minority?
■ confer benefits which can be divided among many or monopolised by the few?

- favour people over machines?
- maximise economic conservation or waste?
- favour the reversible over the irreversible so that undesirable impacts can be removed?

An IT strategy developed from this starting point might look very different from that derived from the usual profit driven basis.

This collection of views challenges the way in which the developed countries and their organisations exploit IT in the name of efficiency and effectiveness. The overall message is social justice and a rejection of any application of IT which undermines and worsens the human condition. For those of you wishing to read more about this topic the December 1996 edition of the New Internationalist is a good starting point. Some of the articles from this edition can be accessed from their website (http://newint.org/issues/1996/12/01 accessed 22/10/2021).

6.3 Understanding in the Community [1999]*

This section was first published as: Rogerson, S. & Begg, M (1999) ETHIcol—Understanding in the Community. IMIS Journal, Vol 9 No 4, pp. 24–25. Copyright © Simon Rogerson & Mohamed Begg.

One of the greatest changes caused by Information and Communication Technology (ICT) is that people around the world communicate regularly sharing their information, ideas, hopes and aspirations. This global village is a rich community comprising people from different faiths, cultures, nations, and ethnic origins. This edition of ETHIcol considers some of the broader community issues specifically regarding Muslims. Why is this important for ICT professionals? ICT professionals must never lose sight of the fact that they are members of a wider community first and foremost, and that membership of the profession is secondary. Understanding the community and individuals within that community will help the ICT professional uphold his or her community obligations.

It is important to appreciate that a large number of people belonging to the Islamic faith are now citizens of the UK, the rest of Europe and the US. According to figures quoted by the Union of Muslim Organisations in UK & Eire, there are about 2 million Muslims in the UK, about 3.5 million in the rest of Europe and 5 million in the US. These figures include a substantial number of indigenous citizens who have embraced Islam as their faith. It has taken over 30 years for these Muslim communities to mature and settle within a Western environment. Community work has been most difficult due to the range of ethnic origins of the Muslims even though the majority appear to be from the Indian sub-continent and the Arab world.

Economically it appears that the Muslim community in the US is much healthier than the UK and the rest of Europe. One reason for this is that the majority of the Muslims who emigrated to the US were educated and professional people. Consequently, there is a large number of Muslim doctors, lawyers and other professionals in the US. They have been exposed to the computer revolution much longer than the Muslims in UK and the rest of Europe, the majority of whom were economic migrants and have only recently started feeling the effects of the computer revolution. Thus, the Muslim community and the associated community workers in the UK and the rest of Europe can learn much from their counterparts in the US who appear to utilise ICT widely and beneficially in both religious work and community work. In the US the application of ICT seems to have successfully crossed the cultural and religious barriers still being felt in the wider Muslim world, where, for example, the general public still does not have access to the internet in Saudi Arabia due to religious arguments.

Some interesting details regarding how the Muslim community in the US organises itself have been made collected when attending the second International Islamic Peace Conference in Washington in August 1998 and the Annual Conference of the Muslim Social Scientists in US held in Chicago in November 1998. Almost all the Islamic organisations have their own websites and majority of their members have email addresses. Every Muslim organisation in the US advertises its activities and functions on the internet. This has enabled most organisations to hold three or four annual week-end conventions. There are, therefore, many conventions to choose from each year.

The usual format is that an organisation advertises its week-end conventions on the internet specifying the dates and venue. Members and visitors, who may be more than a thousand miles away, then make their bookings and arrange their accommodation at the hotel through the internet. Most of the members, families and visitors arrive on a Friday night having taken their flights from different cities. The convention usually starts with a dinner on Friday night with perhaps a speech followed by prayers. The next two days (Saturday and Sunday) are spent at the hotel venue participating in a programme of events, speeches, prayers, lunches, dinners and very often there is also a bazaar where anything can be bought from audio cassettes to expensive outfits to even Islamic computer programs.

Furthermore, in the US computers are used extensively not only by the adults but also by children who learn most of the Islamic rituals and knowledge such as the ALIM, AL-USTADH and AL-QARI through computer programs. A vast wealth of Islamic information is now available through computer facilities. Supplementary teaching has been revolutionised by using these facilities.

It is clear that the Muslim community in the UK and the rest of Europe could similarly improve its communications through the use of the internet facilities. Indeed, these facilities would enable the participation of a vast cross-section of the Muslim community and possibly many of the social and religious issues would then be tackled more effectively. The use of ICT in general could benefit Muslim children with, for example, Islamic computer programs being available locally which would enhance the Madrassa teaching at the supplementary school.

The economic demands of being able to only take unskilled factory work and the religious demands of obtaining scarcely available Halal meat are indicative of the first and second generation emigrants' demise in the UK and the rest of Europe. Therefore, it is hardly surprising that ICT literacy was not prevalent. However, the next generation cannot afford to be ICT illiterate as this technology is progressing and developing at an unprecedented rate and ICT skills are now a mandatory life skill alongside numeracy and literacy. Indeed, the Imams too will have to be ICT literate if they are to function to their full capacity. In addition to giving answers to questions in person, they ought to be able to reply to email questions from people who may be at the other side of the world. Islam is portrayed as a global religion, it will therefore demand global Imams—only the new technologies can provide such a capacity when positively used and without disregarding the social responsibility issues that have arisen through the computing revolution.

It is clear from this account of the Muslim community that ICT provides many positive opportunities in a context that most computer professionals might not have thought about. It is indicative of the overall goal to realise a democratic and empowering technology rather than an enslaving or debilitating one—a goal that all computer professionals should subscribe to.

6.4 Internet Ethics are Not Optional at Business or at Home [2000]*

This section is based upon the 5th IFIP-WG9.2 Namur Award Lecture, Internet ethics are not optional at business or at home, delivered by Simon Rogerson in Namur, Belgium on 14 January 2000. Copyright © Simon Rogerson.

6.4.1 The Internet

The development of an expanding set of international information networks, known as the internet, has been one of the most influential applications of computers and telecommunications. This phenomenon has led to millions of people being interconnected and this will grow to over 200 million by the year 2000. It has evolved from a closed world of specialists and experts to a common and commercial universe open to the general public. The networks of the virtual society offer exceptional possibilities for exchanging information and acquiring knowledge, and provide new opportunities for growth and job creation. However, at the same time, they conceal risks to human rights and alter the infrastructure of traditional public and private operations.

6.4.2 The Problems

Johnson (1997) explains that the potential benefit of the internet is being devalued by antisocial behaviour including unauthorised access, theft of electronic property, launching of viruses, racism and harassment. These have raised new ethical, cultural, economic and legal questions which have led many to consider the feasibility and desirability of regulation in this area. Similarly, it is questionable whether technological counter measures will be very effective either. The absence of effective formal legal or technological controls presents grave dangers to the virtual society. Consequently, there is a need to rely heavily upon our ethical standards.

6.4.3 Internet Ethics

Johnson (1997) suggests that there are three general ethical principles that promote acceptable behaviour in the virtual society:

■ Know the rules of the on-line forums being used and adhere to them.
■ Respect the privacy and property rights of others and if in doubt assume both are expected.
■ Do not deceive, defame, or harass others.

Such principles appear to be built upon core values which most if not all humans would subscribe to. Moor (1998) suggests that such core values are life, happiness, ability, freedom, knowledge, resources and security. However, Moor explains that,

To say that we share the core values is only the first step in the argument toward grounding ethical judgements. The most evil villain and the most corrupt society will exhibit core human values on an individual basis. Possessing core human values is a sign of being rational but not a sufficient condition for being ethical. To adopt the ethical point of view one must respect others and their core values. . . . If we respect the core values of everyone, then we have some standards by which to evaluate actions and policies.

This appears to be an effective conceptual way to address the ethical issues related to the internet.

6.4.4 Business

The outcome of not subscribing to such principles is likely to result in chaos overwhelming democratic dialogue, absolute freedom overwhelming responsibility and accountability, and emotions triumphing over reason (Badaracco & Useemm, 1997). For businesses operating in the virtual society these principles can be expanded into a number of explicit actions such as:

- Establish an electronic mail policy that forbids forgery of electronic mail messages, tampering with the email of other users, sending harassing, obscene or other threatening email and sending unsolicited junk mail and chain letters.
- Clarify the responsibilities of those involved in providing information. The publisher is the producer of the on-line information and is, in the main, responsible for that information. The producer therefore must be identifiable at all times. It is arguable that people can only be held responsible for things they can control. Access providers take this line refuting responsibility for the wrong-doings of users, because their contribution as access providers is purely technical.
- Develop international co-operation that will encourage the creation of common descriptions for internet services, transparency which would benefit users, and respect for trademarks, and increase the possibility of global access to the internet on demand.
- Encourage the development of legitimate electronic commerce that upholds consumer protection through the use of standard contracts and technical mechanisms. Many issues need to be taken into account including the validity of the electronic signature, the solvency of the purchaser, the legitimacy of the vendor, the security of the transaction and the payment of appropriate sales taxes.
- Establish a body for handling customer complaints and reviewing internet activity of the organisation. This body would be in contact with public and private international groups that are competent in internet affairs.
- Train employees in ethical internet practice and promote internet awareness within the wider community.
- Promote a greater equality of access to the internet through multilingual and multicultural support. This counters the current situation where the internet is an Anglo-Saxon network with 80% of servers being North American and 90% of exchanges taking place in English.

Such actions provide the means for self-regulation that would combat the regulatory flux surrounding the internet and might lead to effective policies regarding transparency, responsibility and respect of appropriate legal frameworks.

6.4.5 Home

The home will be the physical location of the virtual society. Many of the organisations with which people interact have used computer technology to provide new forms of interaction that can take place from the home. Information has quickly become one of the most valuable commodities of our society. Information in the future is likely to take many different forms. Taylor (1995) suggests that information will be domesticated as utility networks reach the home of most people and information appliances become cheaper, intuitive to use and interactive. He suggests the types of services and products that might bring about this domestication. They include tele-healthcare, tele-education, entertainment and media, real time information services, electronic publishing, digital imaging and photography, virtual zoos and wildlife experience, virtual reality experiences, and tele-travel. In a short space of time most of Taylor's list has become a reality. It is a total information experience. Indeed, *E.everything* has arrived.

We must be wary of the potential dangers. For example, children can be exploited to reveal personal information about themselves, or they might go one expensive internet shopping sprees or they might down load unsuitable material such as pornographic picture or they might become emotional upset from inadvertently visiting a stealth site (Silverthorne, 1999). Certain practical measures can be used such as filters, locating computers in a family space and restricting the amount of time children can be on-line each day but what is vital is sound moral education. As Karen Jaffe, executive director of KidsNet, a non-profit organisation devoted to children and the media explains, "The most important thing is having conversations with your children about your own values. You can have the V-chip and filters, but if your child doesn't have a sense of what is appropriate as defined by the values of the family or school, then none of this matters."

A second example concern healthcare. The growing moves to create electronic patient records covering a patient's complete medical history from the cradle to the grave is a good illustration of why we must be extremely careful in the way information is created and distributed particularly over the internet. Certainly the electronic patient record allows providers, patients and payers to interact more efficiently and in life-enhancing ways. It offers new methods of storing, manipulating and communicating medical information which are more powerful and flexible than paper-based systems and can accommodate processing of non-textual medical information such as images, sound, video and tactile sense. There are, however, potential problems. The electronic patient record will hold a complete profile of the individual comprising personal and medical details. Access to this information much be carefully controlled ensuring such access is limited to only the relevant and authorised portion of the information. Since medical details contain some sensitive information such as past drug use or genetic predisposition to various diseases it is important to keep this information truly private. There will always be tension and trade-off between the need-to-know and the right to confidentiality. Misdiagnoses are quite rare, but far from unheard of. Procedures must be in place to ensure that once an error is identified the electronic patient record is corrected and all points of distribution informed of the error. Finally, inaccurate data input can be potentially life threatening in this application. It is particularly difficult to correct inaccurate data given the global distribution via for example, the internet, of this information to primary and secondary healthcare providers.

6.4.6 Conclusion

The internet will change society. We must be aware of the potential benefits and dangers and be prepared to challenge any antisocial activity. Therefore, internet ethics are not optional, they must become a way of virtual life at work and in the home. Only then will we reap the benefits that this amazing technology offers.

6.5 Beware of False Gods [2000]*

*This section was first published as: Rogerson, S. & Thimbleby, H. (2000) ETHIcol—Beware of False Gods. IMIS Journal, Vol 10 No 2, pp. 26–27. Copyright © Simon Rogerson & Harold Thimbleby.

This edition of ETHIcol draws upon a talk given by Professor Harold Thimbleby at the Worshipful Company of Information Technologists" third Colloquium on the Ethical & Spiritual Implications of the Internet held on 17 February, 2000 at the House of Lords.

The internet is probably the largest single thing mankind has ever built. We must come to terms with its great size, its multinational, multicultural diversity, and how it impacts on the way we live now and in the future. Many people argue that the internet, like all technology, is neutral. This view is wrong. By denying ethics has anything to do with technology, we do not stop to think about whether or not technological progress is good; it just has to be followed.

The internet neutrality argument is analogous to how the pro-gun lobby in the US argues that guns do not kill people, but people do, and that therefore there is nothing bad about guns themselves. To many, this argument seems to be evasive: clearly, guns and gun making have a serious influence on what damage people can do to each other. Yes, people are responsible for what happens to each other, but guns make some possible courses of action more likely, and more extreme as evidenced by the recent horrific events at an American school. It is true that children do things with guns but we would not like to hold them responsible. Thus, the availability of guns to children is an ethical issue, closely related to the design, manufacture and distribution of guns. In short, technology, albeit guns or computer networks, has an ethical impact on people.

The analogy with guns may be overstated. Guns are meant, at least sometimes, to kill people, whereas computers typically are meant to help people in their work and lives. Nevertheless, we are dealing with a powerful technology and we must, while acknowledging our excitement with it, try to assess the risks clearly. In particular, we need to understand how we fail to appreciate the issues objectively because of the way computers and networks affect our thinking.

We hear a lot of success stories about the internet. Indeed, the media usually picks successes to tell us about. Our senses are bombarded with Net success. We do not hear much about failures, even though there are probably far more of them. One reason is because the internet is a communications medium, and when it stops working, communication stops, so we cannot hear about it. Thus, our thinking is biased since we only see and hear about lots of successes on the internet because the medium is filtering out failures.

Consider the much-heralded internet banking. If you try to pay your credit card off using this facility it can take a very long time to complete the transaction sometimes involving several passwords of many digits. The bank does not see the time it takes. They have no reason to make it easier; in fact, while you are spending time on-line, they are advertising at you. The evolution to internet banking has seen a transfer from a reciprocal world of one-to-one relationships in local banks, to an impersonal one that off-loads work onto customers, in the name of choice or whatever, but which increases profit for central operations.

Computers are much more than passive building blocks that are used to build, for example, the internet. They are practically alive. In many ways computers latch into our psyches, and meet, or almost meet, some of our deepest needs. This is rather like idolatry where, instead of worshipping God, we worship some false god, something that is not a god at all. We want it to rain tomorrow, so we sacrifice a lamb on the altar of the god. But tomorrow maybe it still does not rain, and the priest says we should have sacrificed a bigger lamb or more of them. Somehow the failure of the god becomes our failure, and sure enough, next day there we are buying more sheep to sacrifice.

This is exactly what goes on with computers. First, they are promoted as perfect things. Then, when we discover they do not work, we think it is our fault, and the solution is for us to learn more about the right incantations, and when that does not work, rather than buying sheep we buy more RAM or get a newer even more perfect computer.

While we continue to expect computers to have the omniscient powers of small gods, we are going to be disappointed, and probably pass more and more restrictive laws in the false hope that profound human problems can be solved by machines. When today's computers fail, we are still attached to the idea that tomorrow's will be better. Every year computers are presented as solving our problems. However, the following year these computers are obsolete and we discover that the idol computer does not work. The landfill sites, in the UK are already burying over a million tons of electronics each year. This is idolatry that is not sustainable, and with an EU directive (WEEE, "Waste from Electrical and Electronic Equipment") that comes into force around 2003, it will not be legal either. Those millions of tons of electronic rubbish represent a lot of fooled people.

While we are blind to the ethical and spiritual issues, we will continue to follow blindly so-called progress, having no idea where we are progressing to. The question to ask, is not whether technology is good, but how are we going to do good, given that we live in a world where computers and the internet are expected to be, but are not, perfect solutions to anything?

We should encourage a consumer movement that is aware of these sorts of issues, and a system of standards so we can speak in a clear language. Cars used to be promoted as fast and wonderful; the 1960s consumer movement forced them to be safe, and to be better for consumers. Likewise, we now need a social force that does not spend its time focused on the pinnacles of success, but which represents most peoples' ordinary lives. No doubt it would use the best community-building technologies of the internet, but its purpose would be to help us to hope in each other today, rather than hope in the illusion of perpetually fixing today's problems with tomorrow's computers.

6.6 A Week is a Long Time [2004]*

This section was first published as: Rogerson, S. (2004) ETHIcol—A Week is a Long Time. IMIS Journal, Vol 14 No 2, pp. 30–31. Copyright © Simon Rogerson.

It seems the more sophisticated we become in developing and using technology the greater the risk that it will be used in a detrimental way against individuals, organisations or society at large. We seem to be constantly grappling with new concepts that simultaneously pose new opportunities and new problems, or rushing to pass through new legislation to counter some new threat or redefining societal norms for our shrinking technologically-dependant world. Is this paranoia or reality? Consider the evidence. Several major online news services were sampled for the week commencing 15 March 2004 to identify the current challenges that we face from technology. The findings were concerning.

Over 30 news items covered a range of problems and focused on three perspectives: occurring incidents, technological countermeasures and legislation. The news items fall into several broad categories.

6.6.1 Globalisation

A number of articles addressed various aspects of globalisation. Controversy over offshore call centres and technology development in India was reported in one article. Those for and against such moves sadly seem to align with the political stances of the developed and developing world. In a range of articles about viruses, information access and intellectual property theft the underlying message was the extent and speed of impact due to the global nature of new technologies.

6.6.2 Intellectual Property

This continues to be a hot topic. During the week Microsoft was reported as continuing aggressively to protect its trademarks across the world. Meanwhile the European Parliament passed a draft directive on enforcing intellectual property rights. It was reported that, "Campaigners still believe the move will damage activities such as the development of open source software." There were several reports on content theft. There was a spate of content theft from websites central to the activities of several companies and individuals. Peer-to Peer (P2P) file-sharing appeared in a number of reports. UK firms were shown to be lacking in recognising their professional obligations to society by allowing widespread illegal P2P. Parental obligation was under scrutiny in an article which revealed many parents were either ignorant that swapping copyrighted files was illegal, or knew it was illegal but did nothing about it, or did it themselves having learnt from their children. In the entertainment world a movie piracy ring had been discovered through a server containing many pirate motion pictures. Finally, the music industry continues to worry about P2P and the fact it cannot control the internet.

6.6.3 Identity Theft

The growing problem of identity theft seems to be of little concern of many people. For example, it was reported that one in five Britons were willing to give financial details to unsolicited contacts and were not bothered when £500 was taken from their accounts fraudulently.

6.6.4 Viruses and Hacking

Computer viruses and malware are the scourge of the modern global society. During the week there were several reports on the variants of Bagle and other viruses such as MyDoom and Netsky. The Trojan horse, Phatbot was wreaking havoc as computers continued to be connected to a P2P network of infected machines.

There was growing pressure to address Bluetooth security problems in mobile phones which allowed hacking into the personal data held on mobile phones. But phone manufacturers view the problem as relatively minor. Symantec's biannual Internet Security report confirmed growing severity of hacking incidents. One article reported that work had started on threat and vulnerability tests to ensure that the National Identity Register database, which will form the basis of the UK's controversial ID card scheme, is secure from hacker attacks.

Countermeasures were the subject of several articles. A new antivirus chip was reported in one, another announced that, "A public inquiry has been launched into the Computer Misuse Act (CMA), which could spell the end for current loopholes in the law," whilst one article Recognising that hacking cannot be prevented extolled the virtues of taking out adequate insurance cover.

6.6.5 Junk Mail and Spamming

China was reported as the second largest target for spam with one third of all emails being junk mail. Generally, the deluge of junk mail has led to frenetic anti-spam software development. The latest types were reported as using complex Bayesian techniques to weed out spam from legitimate mail.

6.6.6 Information Access and Denial

It was reported that China had closed two internet sites used by many thousands of people because they carried content deemed to be objectionable by the State. A second report explained that TeliaSonera in Sweden had closed down a website of the Islamic group Hamas because it violated the acceptable use policy. By contrast one article focussed on content filters to restrict access to information for school children to combat such things as plagiarism. Legitimacy of information services was the subject of a contrasting article on fake escrow sites. These spoof sites are used to dupe consumers.

6.6.7 Surveillance

Warnings were given about the need to be sensitive to employees' needs when implementing and operating surveillance systems. A new software tool to detect paedophiles in internet chatrooms was unveiled. One article gave warnings about how everyday accessories such as mobiles phones were increasingly used for surveillance and that such intelligence gathering was commonly shared and condoned.

6.6.8 Health

Our physical health can be at risk from technology. It was reported that those using multiple workstations were at greater risk of developing RSI-type injuries.

6.6.9 Conclusion

In one week the news has carried many disturbing stories about information and communication technology. Each issue has a flip side such as these:

- Freedom of expression—*Censorship by the state*
- Global reach—*Local impact*
- Prevention of computer misuse—*Cure from outcomes of computer misuse*
- Individual ownership—*Community share*
- Information—*Junk*
- Surveillance—*Freedom*

We need to have the ability, tools, and confidence to address such issues. A week may be a long time in computer ethics but if we do not address such issues the consequences will resonate for lifetimes.

Note: News services accessed as part of the research for this article: Financial Times—news. ft.com/technology; Guardian Unlimited Online—www.guardian.co.uk/online/; Silcon.com—www.silicon.com; Tech—Reuters—www.reuters.com; VNUNET—www.vnunet.com; and ZDNET—news.zdnet.co.uk

6.7 Student Rights [2004]*

This section was first published as: Rogerson, S. (2004) ETHIcol—Student Rights. IMIS Journal, Vol 14 No 3, p. 32. Copyright © Simon Rogerson.

Some years ago the current UK government, as part of its policy to get the country "wired," promoted the idea that we should all have an accessible electronic curriculum vitae from the cradle to the grave. The argument was that it would provide benefit for every citizen allowing others to access an "official" account of our achievements thus enabling us to seize career opportunities as they arose.

One of the important issues about this idea is the status of the personal data held on such a record. It seems that such data would become a constituent part of public records. Access to such data then potentially becomes easier and widespread. Richard Gellman (2001) reviewing public record use in the United States suggests that a number of measures should be considered to ensure a reasonable balance between public access and individual privacy. These include:

- Not allowing public access to the certain records
- Limiting private uses of the records.
- Allowing searches for individual records but not disclosure of an entire database.
- Limiting disclosure of records in digital format.
- Giving data subjects more choice about disclosures and uses.

However, the problem remains that with electronic records it is virtually impossible to modify or delete all occurrences of a particular record identified as incorrect or expired. Expunging of data is thus a meaningless concept in the modern digital world.

Government thinking has set the way forward. Today we are experiencing the arrival of electronic patient records and the recently announced national identity card. Will there be no end to this digitization? It seems not. Tony Tysome (2004) writing in The Times Higher Education Supplement discusses a £10million project, by the Joint Information Systems Committee (JISC), that would enable every student in the UK to transfer and update an e-portfolio or record of achievement as they moved through education and into work.

Tysome writes that,

> Two pilot projects, one in the southwest and the other in Northern Ireland, have made good progress in linking data systems in further and higher education and are now turning to schools, but the scope of their work has been limited by data-protection constraints. They have stayed inside the law by restricting ownership of data to institutions and securing students' permission to gather the information when they register. However, this means that access to data becomes "segmented." Bryan Vines, head of project and partnership development at North Devon College, which has been taking part in the Shell pilot project coordinated by Plymouth University, says, "The student can see the whole record, but the institutions can see only their part of the data, unless the individual gives them permission to view all of it." . . . Terry Rourke, the project's manager, says this is fine on a small scale but will not work if the scheme expands. "You could end up with too many data controllers. What we are looking for is a data controller operating at a regional and then at a national level."

The worry seems to be by those involved with the project of how to maintain legal compliance whilst enabling free movement of student data amongst various organisations. There seems to be little recognition of students' rights and that such laws are in place to protect those rights. Looking for ways around the law as implied in the Tysome's report suggests an unprofessional approach.

The Overview of Human Rights in FE & HE by the JISC Legal Information Service states that

> As public authorities Further and Higher Education institutions must now act to
> ensure respect for and protection of the human rights of its staff, employees and stu-
> dents as well as any other individuals that it interacts with. FE and HE institutions
> must not infringe these rights by their own actions or inactions. In that sense the leg-
> islation obliges educational institutions to be proactive in protecting the human rights
> of students and staff. It is no longer enough to not impinge upon their human rights,
> a strategy of safeguarding these rights is required.

It seems very strange therefore that Malcolm Read, JISC chief executive was reported, by
Tony Tysome, as insisting that the moral arguments about electronic student records are not the
concern of the JISC project and that, "Questions like that are not for us to worry about." How can
he abdicate his responsibility in this way?

Not everything is good news. There are bits about our past that we would wish to forget,
demote in relative importance or simply remove for the public view. Is failure or misdemeanour
going to electronically shadow us for the rest of our lives with the advent of electronic records such
as the electronic student record? Once a failure always seen as a failure is contrary to the promotion
of a caring society willing to offer second chances and opposed to preconceived opinion.

ICT strategists and developers such as those at JISC should be mindful of this and not hide
behind technological bushes when difficult moral questions about ICT arise.

6.8 A Global Phenomenon? [2005]*

*This section was first published as: Rogerson, S. (2005) ETHIcol—A Global Phenomenon. IMIS
Journal, Vol 15 No 2, pp. 33–34. Copyright © Simon Rogerson.*

The application of information and communication technologies (ICT) has for long time been
viewed as a global phenomenon that affects all of our lives sometimes positively and sometimes
negatively. This billing has been promoted by a series of opposites such as: haves or have-nots,
information rich or information poor, old or young, developed country or developing country,
online or offline, user or developer, computer literate or computer illiterate, local or global. A
recent trawl of the world's press sheds some light on whether this is a figment or a reality.

In February 2005 The Sunday Times in Malta carried a story by Leo Brincat, the main
Opposition spokesman on Foreign Affairs and IT, about the fluctuations in ICT demand around
the world. He suggested that, "One consequence of the 2001–2002 ICT downturn was that
countries that specialised in ICTs became even more specialised. In fact the global rationalisation
of production has actually led to countries specialising in even smaller ranges of products and
services." This has led to a difficult challenge for the small island of Malta in that this specialist
expansion is "driven by the need for market access, growth, economies of scale and access to skills
and technology." How Malta can continue to compete internationally in ICT remains to be seen.

There are many examples of where ICT has exacerbated extremes. A recent controversy of
alleged internet censorship in Tunisia was reported in allAfrica.com on 4 March 2005. A spokes-
person from Reporters sans Frontières said,

> [Tunisian] President Ben Ali believes that the fact that the United Nations agreed to
> hold a summit on the internet in his country means the international community

approves of his policy in this field. We believe that, on the contrary, the internet model advocated by Tunisia, combining censorship with a crackdown [on dissidents], should be condemned by countries that care about freedom of expression.

The article listed 23 websites that had been censored by officials of the Tunisian President. Freedom of speech will continue to be a major issue as ICT impacts increase globally. A second example was reported by The Age in Australia on 11 March 2005. "Australian recording industry investigators have raided an internet service provider suspected of having used high-speed file-swapping technology to allow the pirating of hundreds of thousands of songs and video clips." The recording industry claimed the ISP was allowing its users to exchange illegally billions of copyrighted music files each month. The ISP's defence is that it is not responsible for the actions of people using their service and software. Such intellectual property conflicts will increase as the world shrinks through ICT dependency.

ICT applications abound. A wonderful example appeared in Thailand's The Nation on 18 February 2005.

> To provide equal access to information, the Thailand Association of the Blind has developed Daisy books as well as the TAB Player, a program to play back Thai-language books for the blind. Short for Digital Assistant Information System, Daisy is an electronic book based on a universal design standard to allow equal access to information for all. Daisy books are based on multimedia formats that combine pictures, sound and text.

Biometrics has been at the application forefront for some time. On 8 March 2005 The Strait Times in Singapore carried a feature on the widespread use of biometrics. The incessant march of biometrics was illustrated by the report that, "Biometrics is creeping to daily living elsewhere too. For example, at Wing Tai Holdings, biometric solutions have been integrated with state-of-the-art condominiums. The biometric systems provide access to individual units and common areas like the gym, lobbies and function rooms." However, there are international concerns about the integrity of biometrics. The 17 February edition of The Economist ran a feature on the use of biometrics in passports. It challenged the integrity of the technical specification of passport biometrics by the International Civil Aviation Organisation (ICAO), a UN agency. It stated that,

> Passport chips are deliberately designed for clandestine remote reading. The ICAO specification refers quite openly to the idea of a "walk-through" inspection with the person concerned "possibly being unaware of the operation." The lack of encryption is also deliberate, both to promote international interoperability and to encourage airlines, hotels and banks to join in.

The dangers of increasing criminals and terrorist threats seem inevitable by such technical specifications.

The press continues to carry many stories about the digital divide. On 25 February New.com.au in Australia reported claims by the World Bank that the digital divide is rapidly diminishing. The World Bank's research had found that 50% of the world's population had access to a fixed-line telephone and 77% to a mobile network. It found there were 59 million fixed-line or mobile phones in Africa in 2002. The World Bank concluded the digital divide was on the wane, "Africa is part of a worldwide trend of rapid rollout. . . . This applies to countries rich and poor, reformed or not, African, Asian, European and Latin American." But not everyone agrees with this. The BT study of the UK, *The Digital Divide in 2025*, found that 51%, or the 24 million adults in the UK are excluded from ICT facilities. It forecasts that "23 million people will remain at risk of digital exclusion in the Britain of 2025." Now consider these examples of a digital divide still very much in existence.

The New Vision on 8 March 2005 in Uganda described the work of an NGO called Uconnect which

> imports used computers from Europe and USA, revamps them and supplies them to schools and organisations. The organisation also helps them to network their computer labs and to get internet connections. Uconnect project works to make the spread of internet to remote areas sustainable, scalable and reproducible.

In contrast the Trinidad and Tobago Express on 2 March 2005 ran a story about Microsoft donating software to improve access to ICT. The report explained that "thousands of Trinidad and Tobago's citizens lack the access and, more importantly, the skills they need to participate in the new information-based global economy to realize their full potential. Microsoft is determined to dramatically improve those statistics. Microsoft believes that publicly accessible gathering places, whether they take the form of libraries, meeting rooms, schools, or community centres, represent prime locations where people can go beyond merely having access to technology and can acquire the skills to use it effectively to help themselves and their communities."

These two examples illustrate an international effort attempting to address a real digital divide. This is analogous to the world's approach to relief aid. The latter provides tools to grow and harvest crops whilst the former provides tools to farm and produce information. Both are focused on sustainability and self-sufficiency. But more is needed to address issues outlined in the Kenya Times on 11 March 2005 by Raphael Tuju, the Minister for Information and Communication. He explained that,

> one of the challenges in the region was on human resource development as a large proportion of the population have limited access to the Information Society. The shortage of ICT workers is further worsened by global migration of skilled workers and insufficient pipe lining of ICT education to potential students from primary to tertiary levels of education.

Drawing upon examples from Australia, Kenya, Malta, Singapore, Thailand, Trinidad and Tobago, Tunisia, Uganda, and the UK, this brief review has shown that ICT has diverse impacts across the globe. Consequently, ICT professionals have enormous influence and obligations in this new order. But do we have the wherewithal and inclination to rise to that global challenge?

6.9 Safety on the 'net [2007]*

*This section was first published as: Rogerson, S. (2007) ETHIcol—Safety on the 'net. IMIS Journal, Vol 17 No 2, pp. 31–32. Copyright © Simon Rogerson.

In an ever-increasing online world it seems we need more and more help to ensure we can live out a peaceful virtual existence and not suffer from criminal attack, hooliganism, intimidation or general intrusion. For example, as reported in the Guardian 3 February, mass-marketed scams are costing adult victims £3.5bn a year, according to new figures from the Office of Fair Trading (OFT). Mike Haley, head of Scambusters at the OFT, was reported as saying,

> Scammers are finding more ruthless and sophisticated ways to exploit modern tools such as the internet, email or text messaging. It is important to dispel the myths that scams are victimless crimes and involve only small amounts of money. These scams deeply affect individuals and families, leading to debt, depression, and even suicide. Though anybody can be conned, it is always the vulnerable who suffer the most.

Children are similarly at risk. For example, "Bullying by mobile phone, email or over the internet is a growing problem, according to a survey by the Teacher Support Network and the Association of Teachers and Lecturers" (Guardian, 12 February).

Whilst laws and regulations can provide a framework to deal with perpetrators, it is we, the virtual citizens, who must practice self defence. That is why services such as Get Safe Online, the UK's first national internet security awareness campaign, are very useful. It is a joint initiative between HM Government, the Serious Organised Crime Agency (SOCA) and BT, eBay.co.uk, HSBC, Microsoft and SecureTrading, and aims to "raise awareness and to help educate the general public and small businesses about the dangers of e-crime and how to use the internet safely and securely, primarily through the campaign website www.getsafeonline.org."

This service provides useful advice on living and working virtually. This advice is provided in three broad categories: Protect your PC—block viruses, hackers and other threats; Protect yourself—protect your family, money, and privacy; and Protect your business—extra advice for small businesses.

For example, a recent addition to the website addresses the issue of online dating. It suggests that there are serious risks associated with meeting people online. These include:

- Personal safety when meeting someone you met online.
- Stalking and harassment.
- Meeting people who shouldn't be dating online.
- Dating sites being used as vehicles for spam, selling or fraud.

It is suggested that you choose your dating forum carefully. "For example, you should look for a site that will protect your anonymity until you choose to reveal personal information. Also look for a site that will enforce its policies against inappropriate use." It is further suggested that you protect your privacy.

> You are in control of what happens. Don't let anyone pressure you into giving away more information than you want to. You wouldn't give your phone number to every stranger on the street. Similarly, don't post personal information, such as phone numbers, in public places on the internet.

Such advice seems obvious but there are many internet residents who fail to realise that this virtual world can be as dangerous as the real one and simply leave their common sense at the keyboard.

For small businesses there is sound advice on how to establish a secure internet presence through effective staffing and infrastructure policies. Throughout there are salutatory real case studies of what can go wrong and the resulting devastating consequences. The danger of wireless networks is illustrated by the following example.

> His small office contained three desks and three computers, all networked together and connected to a broadband internet connection using a wireless network.
>
> Two of his staff were out with clients and he was alone in the office with a friend who helped them out with IT problems.
>
> The internet connection had been getting slower and slower and he couldn't figure out why.
>
> The computer expert quickly found the problem. Trying to secure the wireless network, he had looked up who was connected to the wireless network.
>
> There were five users connected. But only three PCs in the office and only one of them was switched on. His neighbours in the office building and in the residential street outside were simply freeloading on his connection. He was paying for the broadband and they were blagging it.
>
> The friend quickly locked them out and hoped that none of them had used the connection to hack into one of his friend's computers.
>
> An extensive 2002 survey found that 92% of wireless networks in London had not taken basic steps to restrict access. That's over 4,500 networks that are wide open to people freeloading internet bandwidth or, worse, snooping on private data.

And Get Safe Online seem to practise what they preach. The terms and conditions of use of the website state: . . . reserves the right to remove from and block contributions . . . which

- are likely to disrupt, provoke, attack or offend others
- are racist, sexist, homophobic, sexually explicit, abusive, or otherwise objectionable
- contain swear words or other language likely to offend
- break the law or condone or encourage unlawful activity (this includes breach of copyright or right of confidentiality, defamation, and contempt of court).
- advertise products or services for profit or gain
- are seen to impersonate someone else
- include contact details such as phone numbers, postal or email addresses
- are written in anything other than English
- contain links to other websites
- describe or encourage activities which could endanger the safety or well-being of others
- are considered to be "spam," that is posts containing the same, or similar, message posted multiple times

This seems sound advice for anyone operating a website that supports interaction with and between visitors. It is clear that we all need to play our part in ensuring anyone and everybody experiences a peaceful virtual existence.

6.10 Reflections from China [2007]*

This section was first published as: Rogerson, S. & Bynum, T.W. (2007) ETHIcol—Reflections from China. IMIS Journal, Vol 17 No 3, pp. 29–31. Copyright © Simon Rogerson & Terry Bynum.

6.10.1 Introduction

Since its inception in 1995, the ETHICOMP Conference series has provided a forum for scholars and practitioners worldwide to engage in debate and dialogue concerning the social and ethical implications of information and communication technologies. IMIS has always been associated with ETHICOMP. In order to engage with those from other cultures and communities across the world, ETHICOMP 2007 was held at Meiji University in Tokyo, Japan and a new initiative, the ETHICOMP Working Conference 2007, was held at Yunnan University in Kunming, China. In this edition of ETHIcol we reflect upon our experiences in China, which included the two-day ETHICOMP Working Conference in Kunming, lectures and conversations within the Schools of Computer Science and Business at Zhejiang Wanli University in Ningbo, the Department of Philosophy at Peking University and the Chinese Academy of Social Science. Especially important was a conversation we had with Ke Xiang, Master of the Ningbo Seven Pagoda Temple and Member of the Board of Directors of the China Buddhist Association.

6.10.2 Two Initial Perspectives

While discussing ethics and information and communication technologies at the ETHICOMP Working Conference 2007, it became apparent that the "technologist's perspective" in China was remarkably similar to that of technologists in Europe and North America before computer ethics was fully integrated into university curricula. In both cases, technologists, in general, believed that "technology is technology and ethics is ethics, and the two are not related." This narrow view is problematical because it causes technologists and public policy makers, to ignore the wider social and ethical impacts of technology. Issues like social inclusion, intellectual property ownership and individual privacy, for example, can easily be overlooked with such a narrow outlook.

A broader perspective was expressed by the Buddhist Master, Ke Xiang when he used an analogy between a table and the internet. Just as coming to the table brings people together, he explained, so using the internet can have a similar effect, but in a much broader sense. At the same time, he acknowledged that the internet has many ethical pitfalls. He said that it is not technology that should be central in our thinking, but human beings and how they educate their children. If parents and society properly educate children to become ethically responsible and virtuous citizens, their interaction with and use of technology will be socially appropriate.

In our view, this wise advice should be applied to all levels and areas of education. Therefore, computer science courses at universities should not only address the science, but also the social impacts. In particular, we believe it is paramount that computer ethics be embedded within the computer science curriculum. Indeed, this multidisciplinary education of professional engineers, including those involved with computers, was recommended nearly 20 years ago by a Chinese academic. In an oral history of China, Portraits of Ordinary Chinese, edited by Liu Bingwen and Xiong Lei (1990) there is a fascinating insight by Xie Xialing who "now a lecturer in philosophy at the prestigious Fudan University in Shanghai, was once a 'Red Guard' in the early days of the Cultural Revolution" He became disillusioned and outspoken which led to him being imprisoned and then exiled as a "class enemy." "Originally he majored in engineering, but his experiences as a Red Guard changed his academic orientation." He explains that, "China needs more engineers than philosophers, but it won't do if one knows everything about machines, but nothing about Plato or Kant, literature, or the Cultural Revolutions." This type of approach presents a great challenge to university curriculum leaders and government policy makers in ensuring adequate time and resources for computer ethics instruction. We were heartened by the fact that a number of

Chinese academics were very interested in experimenting with the inclusion of computer ethics within their courses.

6.10.3 Computer Ethics Issues in China

The urgent need for this curriculum update is illustrated by the fact that we discovered, by reading just a few issues of the newspaper China Daily, that the Chinese people already are experiencing a wide range of complex computer ethics issues. The following examples clearly illustrate this point:

1. In the China Business Weekly, 2–8 April 2007, page 9, Li Jing wrote, "China is considering ways to establish its own intellectual property right (IPR) policies and standards for information and communication technology (ICT), to strike a balance between protecting patent-holder rights and encouraging interoperability among "islands" of technologies.

   "Though standards and IPR are both important for technological progress and improving social welfare, there is an inherent conflict between the two when a patent is integrated within standards, says Zhou Baoxin, Secretary General of China Communications Standards Association (CCSA), which has been drafting IPR strategies for standardization since 2003."

2. In the China Daily, 31 March – 1 April 2007, page 9 (quoting the Oriental Daily), it was reported that "The Office of the Privacy Commissioner for Personal Data is investigating a blacklist of more than 100 foreign maids that was posted on a Hong Kong parenting website.

   "A post entitled 'Come in if you want to blacklist a maid' appeared on a forum of the popular website Baby Kingdom and has been viewed more than 15,000 times. The names, photos and passport numbers of the 100 maids who had been accused of stealing, beating children, lying or being lazy had been added to the thread.

   "The operator of Baby Kingdom has denied responsibility, saying it would be impossible to monitor every one of the 10,000-plus messages left on the website's forums daily."

3. In the China Daily, 10 April 2007, page 1, Wang Xing reported that, "The government yesterday issued a regulation, which takes effect on 15 April, demanding online operators set up a 'game fatigue system' that encourages players under 18 to play less that 3 hours a day.

   Online gamers will also be required to register using real names and identity card numbers to indicate if they are younger that 18.

   Experts said the move reflects government fears over the social impact of popular online games, which have been blamed for the rising numbers of school children playing truant or even committing crimes."

4. In the China Daily, 12 April 2007, page 10, Li Xing discusses Tim O'Reilly's proposal for a code of conduct for bloggers. She writes, "I believe that such a code is necessary in China, especially to avoid the spread of verbal violence on the internet.

   We Chinese should be familiar with what havoc verbal violence can wreak on society.

   Some people may argue that a code of conduct not only deprives netizens of their freedom of expression but also discourages criticism, which society needs for healthy development.

   Indeed, criticism is like doses of bitter medicine to help treat the ills and wrongs in society or offer advice for better creative arts and other works.

   But viciousness, slander and disrespect for people's individual choices and private lives are not criticism. They are poisons that harm the freedom of individuals and society harmony, whether they appear on the internet or in society."

5. In the China Daily, 31 March – 1 April 2007, page 4, the Beijing-based scholar A Ying wrote that "In today's world, the spread and retrieval of information no longer suffer from the confines of class. The great mass of grassroots citizens receives information almost at the same time as those in other social strata and therefore enjoys the same opportunities and conditions for reaching their own conclusions, hence the rising value and importance of their opinions and viewpoints.

Thanks to the public and anonymous nature of the internet, the political moods of netizens inevitably reflect the overall mood of the society. What the blogs reflect is one side of public opinion, popular culture and social interests, though the practice of attracting visits with sensationalism and low taste is constantly targeted by many netizens. In general, the rise of grassroots web users in China is undoubtedly an important area closest to the world-wide trend in this country's modernisation process and a new subject worth studying and close observation." The "new subject" referred to here is computer ethics.

6.10.3.1 The Digital Divide

The growing dependence of society upon technology means that citizens who already are disadvantaged, for whatever reason, are likely to become even more disadvantaged. This is what Western scholars have named "the digital divide." An excellent example of this in China was described in the China Daily, 2 April 2007, page 1. There Wang Zhuoqiong reported that illiteracy increased by 30 million between 2000 and 2005 despite government initiatives. Wang noted that "The number of illiterates in China accounted for 11.3 percent of the world's total in 2000 . . . and 15.01 percent in 2005." One reason offered to explain why this happened was that farmers can earn more money working as laborers in addition to farming, rather than going to school. This explanation was suggested by "Gao Xuegui, director of the illiteracy eradication office of the basic education department of the Ministry of Education."

Being illiterate makes it difficult or impossible to use new technology. As a result a growing number of illiterate citizens become more disadvantaged as society becomes more dependent upon technology. The illiteracy problem in China, as reported in the China Daily, is likely to lead to a wider and more challenging "digital divide." When dealing with a country as large as China, the problem is more complicated and more difficult to resolve.

6.10.3.2 Concluding Remarks

The previous examples of computer ethics issues in China confirm the need for Chinese technologists, business leaders, government officials and public policy makers, as well as Chinese scholars, to be proactive in developing solutions to the complex, technology-generated social and ethical challenges. We were inspired by our conversation with Buddhist Master Ke Xiang, which clearly demonstrated that the ethical wherewithal and wisdom to deal with computer ethics challenges already exist within Chinese society. Just as the social context in which technology is deployed determines how that technology will be used, so also the wisdom and social values of a community provide the wherewithal to address the computer ethics issues that emerge. It is clear that the great traditions and teachings of Chinese philosophers and thinkers will not only enable China to address technology-generated ethical issues within its borders, they also will enable China to make a significant contribution globally.

6.11 Digital Slavery [2007]*

This section was first published as: Rogerson, S. & Rogerson, A. (2007) ETHIcol—Digital Slavery. IMIS Journal, Vol 17 No 5, pp. 31–32. Copyright © Simon Rogerson & Anne Rogerson.

It is so easy to be taken in by the positive spin surrounding ICT. We are bombarded with the latest declarations about new technological advances which will make our lives easier and more enjoyable, saving us time and money. Technologists, traders, and politicians extol the virtues of new applications and decry any attempt to hold on to existing ways of doing things. More and more of the services and products we consume and the way we interact with public agencies, such as the Inland Revenue, are made available online. Eventually many, if not all, of these services, products and interactions will only be available online, thus forcing us into the virtual world.

In this virtual world we exist through a myriad of personal data and electronic interaction. We are digital beings who live in data repositories and travel along the conduits of data communication. Conduits and repositories are owned by others but the claim that our personal data and electronic interactions are owned by others is tantamount to accepting that we, as digital beings, can be owned by others, albeit in some form of distributed cooperative. With ownership comes the right to use, trade and dispose. Existing legislation such as data protection is concerned with the legitimate use of data items. It does not consider data items to be the organs of a digital being and so is not concerned with the welfare of digital beings protecting them against servitude and slavery.

The Universal Declaration of Human Rights states that "No one shall be held in slavery or servitude; slavery and the slave trade shall be prohibited in all their forms." Today that must include both physical slavery and digital slavery. Two hundred years ago in 1807 the Abolition of the Slave Trade in Britain heralded the beginning of the end of slavery as a legal and legitimate state of being, although it was not until 1833 that the abolition finally occurred with the outlawing of slavery In British territories. The parallels between physical slavery and digital slavery are striking.

The economic imperative which led to the growth of slavery was the need for substitute labour to replace the indigenous population of America wiped out by European diseases coupled with the demand for increasing labour as the new world economy expanded. Similarly, there is an economic imperative fuelling the growth in digital slavery. This is the result of the need to develop the conduits to sustain virtual trading.

Slavery thrived through a combination of power and ignorance. Some of the first slaves to be traded in West Africa were "sold" to European traders by powerful local chiefs, complicit in the subjugation of their own people. The trade continued to flourish partly as a result of cultural differences between the owners and their slaves. The former believing, in some cases, that the latter were almost sub-human, and so their enslavement could be justified in the same way as ownership of any other animal. The same combination of power and ignorance has enabled digital slavery to flourish as well. There remains public ignorance about ICT in terms of its capabilities, limitations and applications. It is this which has enabled personal data items to be considered as proxies rather than organs of data subjects. It is the technologists who have knowledge and with that knowledge comes immense power. It is this power which has resulted in society becoming increasingly dependent upon technology. In these conditions digital slavery can incubate.

The winners and losers in physical slavery are obvious. Those enslaved will lose and be forced to live in a world of oppression and degradation whilst the owners of slaves benefit economically and socially. In digital slavery we are all losers because at some point even so-called owners will themselves be enslaved as they themselves seek to consume products and services which are only provided online.

In 200 years the wheel seems to have turned full circle. Digital slavery exists, unrecognised, unnoticed, unfettered, and, in many cases, unconsciously supported and promoted. Governments, for example, with the creation of biometric identification, have created the shackles of slavery. Digital beings are traded online as owners try to maximise market share and increase profit. Such trading is unacceptable and contravenes human rights. In 2007 it is time to call for its abolition and so mark the end of digital slavery.

6.12 Inclusive ICT [2009]*

This section was first published as: Rogerson, S. (2009) ETHIcol—Inclusive ICT. IMIS Journal, Vol 19 No 1, pp. 33–34. Copyright © Simon Rogerson. Note: An earlier version of this article appeared in Connexion, No 29 pp. 1–2 at www.hitachiforum.eu/Connexion/documents/Issue29.pdf

ICT has always been a powerful change agent. Many have turned to ICT to help resolve the growing concern over social exclusion, and the care of old people and those with disability. It is true that ICT can provide new opportunities for improvement in people's daily lives including work, education, travel, entertainment, healthcare, and independent living. However, there is a risk that, despite its many benefits, ICT could set people apart, create new barriers, and increase social exclusion. Specific attention needs to be given to those groups in society which are at high risk of being excluded, due to a wide variety of reasons such as age, gender, disability, literacy, and culture. How ICT strategy is developed and implemented can mean for disadvantaged people the difference between dependency and autonomous living.

New technology often focuses on maintaining and monitoring the health of older people but technologists must take a broader view. The common misunderstanding that looking after the elderly simply means addressing health and welfare issues needs to be rectified. Older people are more physically and mentally active than previous generations and this should be a key consideration in guiding the development of new services and products for them. Indeed, technology can provide a vital link for older people to family and friends through email, instant messaging or video conferencing, but there are many other potential applications. These could include: virtual travel and tourism (virtual tours of museums and art galleries for example); support to enable older people to continue working, should they wish to; and enhanced media, such as books, films, and music, which take into account the specific needs of older people, such as poor hearing or eyesight.

Therefore, ICT must be used to not only address health and welfare issues but also for social interaction, life-long learning and work. This is because in the Information Society we have an ageing population which is increasingly more physically and mentally active and has a younger outlook than in generations past. If ICT is to play a part in supporting the ageing population it must be financially accessible, usable, useful and transparent. Old design and business paradigms must be challenged. That is starting to happen!

Enter the Flip Video—a pocket video camera which is easy to use and cheap. Gone are complex multiple controls, gone are bulky external chargers, gone are multiple wires to connect to computers and televisions, gone are memory card slots and gone are CD-ROMs of software to manipulate videos on a computer. In their place are a few simple buttons, two AA batteries, a flip-out USB plug, embedded memory to hold sixty minutes of video, and software in the camera which automatically loads on any computer when the camera is plugged into a USB port. So why is this so special? It is because it could fundamentally change the way we view and interact with people and places around us. It could revolutionise, for example, education, entertainment, reporting and communication. It could enable everyone, including older people, to interact using video, rather than text or voice-only, which is intuitively more natural. Design-for-all is evident in this little camera. Jonathan Kaplan, founder of the Flip Video company, Pure Digital Technologies explained, "We want to have software that helps users feel smarter." It is a great example of technology which puts people first. It shows the direction in which we should be moving so that the development and use ICT is empowering rather debilitating and inclusive rather than exclusive.

At the other end of life what is happening for children? There have been many government initiatives around the world to ensure our children can have the opportunity to access and benefit from ICT. Many of these initiatives have floundered primarily because they focused on ICT provision through institutions such as schools and libraries. But this has changed with the advent of low entry computers. Two examples are the Classmate PC, formerly known as Eduwise, which is Intel's low-cost personal computers for children in the developing world and the One Laptop Per Child (OLPC) trade association's Children's Machine (XO) which is for a similar market. Both will operate Linux and so use the free and open-source software environment thus making these computers so much more affordable. These computers provide great education in a rugged industrial design intended for children but the design is not childish. Such initiatives offer a glimpse of accessible fit-for-purpose ICT for all children of the world.

Technologists have a great opportunity to really make a positive difference to people's lives. With some thoughtful design under a strategy of inclusion ICT could become the instrument of equality of opportunity and sustainable quality of life. Think of the job satisfaction from realising that vision.

6.13 That Was The News That Was! [2009]*

*This section was first published as: Rogerson, S. (2009) ETHIcol—That was the news that was! IMIS Journal, Vol 19 No 4, pp. 28–29. Copyright © Simon Rogerson

For this edition of ETHIcol a review of just two newspapers in the UK has been undertaken for one week to see what, if any, are the current ethical issues with ICT. It seems that ICT is a problematic as ever.

6.13.1 Trade Press

Read these seven extracts from Computer Weekly. They reveal that serious ethical issues continue to challenge the industry. The balance between the need to substantiate one's identity and civil liberty continues to be in the news with the announcement of the new ID card. Hacking in various forms is an ongoing issue as ICT pervades our lives and opens up new opportunities for those with criminal or antisocial tendencies. Property theft remains a big issue as intellectual property owners strive to protect their property. ICT's environmental track record has become a hot topic with use of raw materials and energy consumption being challenged. The final story is a good example of how a new technology can change the way we interact with each other—have we ever stopped to consider the implications of this before we embark down a road that cannot be easily changed.

Identity revealed

"The design of the ID cards that British citizens will carry has been unveiled today. Each card will have a photograph of a face and will contain the name, date of birth and signature of the holder. It will hold similar information to that currently contained in the UK passport but will also a feature biometrics, with a photograph and fingerprints stored on a secure electronic chip."

Hacking new media

Apple has failed to issue a security patch for a vulnerability that could allow criminals to take control of iPhones using text messages, say security researchers. Apple was believed to be working on a patch after researchers notified the firm of a vulnerability in the way the iPhone handles text messages. They warned that once exploited, the iPhone flaw could be used to take control of the devices dialling, web browsing and texting functions.

Intelligence Hacking

The website of MI5 was breached by hackers when a search engine related to it was penetrated. The problem was not a direct threat but it is cause for concern, according to security experts. According to a report in the Daily Express the hackers could have put viruses on the computers of people using the website and find out their identities.

The weakest link

The first crack to defeat the anti-piracy mechanism built into Microsoft's forthcoming Windows 7 operating system has been reported three months before general release. Reports of the crack for the Ultimate edition come less than a week after Windows 7 was released to original equipment manufacturers (OEMs), according to Softpedia. The crack is based on a product activation key and certificate believed to have been extracted from an OEM copy of Windows 7 that was leaked on a Chinese web forum. The crack is believed to be limited to the Ultimate edition, but can be used on all machines for both the 32-bit and 64-bit versions of the operating system. And if a user has a retail version of Windows 7 Ultimate, it can be converted to an OEM version with two simple commands, and then activated, according to Ars Technica.

Property theft

Businesses across Europe, the Middle East, and Africa have paid out £6m this year to settle disputes with the Business Software Alliance. The BSA has received £435,000 from UK companies in settlements and payments for getting properly licensed in the first half of this year. Overall, firms across EMEA paid out £2.1m in settlements and £4m in licence costs. The BSA has pursued a policy of taking legal action against companies that break software licensing laws and accepting settlements from those that are prepared to admit wrongdoing.

Environment

Greenpeace protesters demonstrated outside HP's US head office today to highlight the hardware maker's policy on using toxic chemicals in its production processes. HP [said it] supports industry efforts to eliminate BFR and PVC because of potential e-waste issues. HP is a worldwide leader in e-waste recycling. HP has recycled £1bn of electronic products between 1987 and 2007, and has committed to recycling another £1bn worth between 2008 and 2011

New ways of communicating

The UK government has published a guide explaining why and how public sector bodies should use Twitter. The government is joining businesses, including banks, in harnessing Twitter to communicate with customers effectively. The government is encouraging MPs and civil servants to use the microblogging website, which allows users to post 140-character messages in real time.

6.13.2 Popular Press

It is not just the trade press that carries such stories the popular press often has ICT stories with an ethical dimension. The Guardian had several of the stories that ran in Computer Weekly. There were other stories. Read the three extracts for the Guardian in the same week. Catching computer criminals is one thing but bringing them to justice is another. ICT has revolutionised information exchange but sometimes this can be detrimental to the public at large. The file sharing legacy of Napster continues with the debacle over The Pirate Bay.

Hacker extradition

There were emotional scenes outside the high court today after computer hacker Gary McKinnon lost a further attempt to avoid his extradition to America on charges of breaching US military and Nasa computers. The verdict follows the latest in a long line of legal moves by McKinnon, who describes himself as a "UFO enthusiast," after he was accused in 2002 of using his home computer to hack into 97 US military and Nasa computers, causing more than $700,000 in damage according to the US.

Too much information?

The online encyclopedia Wikipedia has become embroiled in a bitter row with psychologists after a Canadian doctor posted answers to controversial tests on the site. The Rorschach test is designed to give psychologists a window into the unconscious mind, but many now fear their patients will try to outwit them by memorising the "right" answers. The row erupted when hospital doctor James Heilman from Saskatchewan posted all ten inkblot plates on Wikipedia alongside the most common responses given to each. Heilman uploaded the images after becoming frustrated by a debate on the website as to whether a single Rorschach inkblot plate should be taken down. "I just wanted to raise the bar," he said. The move brought immediate condemnation from psychologists who signed on to complain that making the tests public renders them useless. "Making images available on the internet will make it obsolete and we will have lost a helpful tool," said one.

File sharing

In yet another legal blow for The Pirate Bay, a court in the Netherlands has ruled that the file sharing site must block its site for Dutch internet users within 10 days or face stiff fines. Failure to comply with the ruling will result in fines of €30,000 a day up to a maximum of €3m for the three founders of the site, Peter Sunde Kolmisoppi, Fredrik Neij and Gottfrid Svartholm Warg. This comes after a dozen movie studios filed a suit seeking to shut down the site this week. They were seeking the injunction after an April ruling by a Swedish court that found the three founders and funder Carl Lundström guilty of helping millions of people download copyrighted material. They were given one-year prison terms and fined 30m kronor (£9.1m).

As can be seen from these news items, ICT remains an ethical melting pot. Professionals have obligations and responsibilities to ensure the melting pot does not spill or that we do not get our fingers burnt.

6.14 Assessing All the Risks [2011]*

*This section was first published as: Rogerson, S. (2011) ETHIcol—Assessing All the Risks. IMIS Journal, Vol 21 No 2, pp. 38–39. Copyright © Simon Rogerson

This month sees the culmination of ETICA, an important FP7 European research project which investigates the ethical issues of emerging ICT applications. One of the final events of the project was to host at the European Parliament "IT for a better future: How to integrate ethics, politics and innovation." The author of ETHIcol addressed the assembly on "Why care about the ethical issues of emerging ICTs" and this is the subject of this edition.

2004 was the year of Web 2.0 and when social media was realised. Social media describes sites and web applications that allow its users to create and share content and to connect with one another. The biggest innovation of 2007 was almost certainly the iPhone. It was almost wholly responsible for renewed interest in mobile web applications and design and caused social media to change the world forever. Most of the readers of ETHIcol have travelled in at least some part of the social media landscape. Today 9.6% of the world's population uses Facebook and shares over 7 billion pieces of content weekly. There are 95 million daily Tweets on Twitter, 35 hours of video uploaded every minute on YouTube and 3000 images uploaded every minute on Flickr.

On 16 March 2011 Viviane Reding, Vice-President of the European Commission and EU Justice Commissioner, delivered and important speech "Your data, your rights: Safeguarding your privacy in a connected world" which reviews the EU Data Protection framework. She laid out a privacy agenda for social networking based on four policy pillars; "right to be forgotten," "transparency," "privacy by default," "protection regardless of data location." This is an important change in how privacy issues are going to be address politically. There has been much discussion about privacy and social networking—it is an important issue and is now under the political spotlight.

However, there is a tendency for this focus on an obvious ethical issue, such as privacy, to obscure any other ethical consideration which indeed might be as important if not more important. Often issues are interrelated which further complicates their identification at first and the manner in which to address them thereafter. For example, during the last 12 months five papers have been published about other ethical issues surrounding social networking.

Batemen, Pike and Butler (2011) focus on public space. They explain that users desire social interaction and disclosing information plays and essential role in such interaction. However, they may not wish to have such information publicly accessible to an unknown audience. Therefore, there exists a psychological tension in social networking users.

Witt (2009) focuses on multiple usage. Social networking sites (SNSs) whilst using for social interaction are used for other things. For example, they are used by companies to explore personality traits when hiring staff and they are used by lawyers to find incriminating information in divorce and child custody cases. There is thus a sense of *Dual Use* where SNSs present a danger to participants or society as a whole if improperly accessed or used.

Strutin (2011) focuses on a different twist to monitoring. The importance of SNSs as a legal penalty is now accepted through case law in the US. A juvenile gang member was sentenced to probation which included having no access to any SNS so minimising the temptation to contact other gang members. This is judicial recognition of SNSs as a communication media that can be monitored. As such SNSs now have a different status within society.

Louch, Mainer and Frketich (2010) look at shaping personality. The teenage experience is one where adolescents shape their identity. Advertisements as part of the SNS experience have an impact on who or what teenage participants become. Targeted marketing of such participants using data profiles is social manipulation and discriminatory.

Finally, Light and McGrath (2010) look at moral agency. They explain that SNSs have moral norms and values embedded in their structure and operation. As such the technology shapes the user experience from the onset. This influences what they concentrate on when using social networking and consequently diminishes their focus on other things such as privacy safeguards.

As can be seen, the ethical landscape of social networking is rich, varied and complex. It is much more than just privacy which clearly is an important issue. It is vital that all existing and emerging ethical issues are thoroughly addressed by researchers, policy makers, politicians, and vendors so that a balanced and comprehensive view of social networking can be achieved.

This example leads to a general point about creating simple lists of issues. In many official documents the phrase *Privacy and Ethics* is often used. This is misleading and inappropriate. Privacy is part of ethics. In using this phrase, it places undue emphasis on a single issue. The very creation of simple lists of ethical issues is problematic. It promotes a culture of compliance. The emphasis is to tick off issues on the list as having been considered and therefore ethical compliance is achieved. What is needed is a culture of ethical sensitivity to be nurtured. A culture where action is based on the knowledge and commitment that such action is the right thing to do rather than on the relief that compliance has been achieved.

Emerging ICTs will throw up complex ethical issues which have to be addressed. If not, at best we will fail to realise the potential benefits for us all, and at worst there will be risky ICT which could severely harm any of us and or the environment. That is why we need to care about the ethical issues of emerging ICTs.

6.15 Academic Publishing in the Information Age [2017]*

*This section was first published as: Rogerson, S. (2017) Academic publishing in the information age—an editor's observations. Journal of Information, Communication and Ethics in Society, Vol 15 No 2, pp. 106–109. Copyright © Emerald Publishing Limited. Reprinted by permission. DOI 10.1108/JICES-03-2017-0017

6.15.1 Introduction

I am the editor and founder of the Journal of Information, Communication and Ethics in Society (JICES) published by Emerald Publishing. The journal focuses on the social and ethical issues related to the planning, development, implementation and use of new media, and Information and Communication Technologies (ICT). The journal was launched in 2003 by Troubador, an independent UK publisher based in Leicester. In January 2007, the journal was acquired by Emerald its current publisher. Since the inception of JICES, publishing; and academic publishing in particular; have changed beyond all recognition. It has been an evolution which has involved everyone; academic author, editor, reviewer, publisher and of course reader. It is a revolution which reflects technological advances and societal acceptance of such advances together with the associated changes in habit, activity and norms.

6.15.2 Journal History

When JICES was launched communication between the two editors and the publisher tended to be local and face-to-face. The journal was produced in a traditional manner resulting in paper-based editions. Communications between editors, publisher and authors moved quickly to email and sometimes teleconferencing, but the paper-based product prevailed. At the time of Emerald becoming the publisher a major technological change took place in the form of ScholarOne Manuscripts, an electronic submission, reviewing and editing system. This has become the current leading journal and peer review tool for academic publishers. Alongside the paper-based editions an electronic version was introduced which was accessible through, for example, university electronic libraries. As electronic publishing and readership access matured so the move to an electronic only version was inevitable. This happened in 2014. Paper-based versions remain available on demand but at additional cost. The complete publishing cycle was now online.

Academic career progression has always been influenced by publication track-record. However, the demand to publish often and quickly seems to have increased. Linked with this is the need to demonstrate academic influence through citation count. These factors have caused a change in academic publishing. There is a move, JICES included, away from volume/issue as the unit of publishing to academic paper as the unit of publishing. This new approach means that as soon as a paper is accepted it can be made available through online access and so journal volume and issue become of secondary importance.

6.15.3 Virtual Space

As can be seen ICT is at the heart of the academic publishing revolution. The online world has become the norm. An extensive virtual network exists in which those involved are all potentially connected. New modes and patterns of interaction and expectation have developed which challenge our social norms and moral integrity. It is within this complex set of relationships that academic publishing exists. Driven by institutional strategy and priority, academics and students undertake research which is written up and submitted for publication. Technology-enabled access to data and literature informs the research and writing processes. Choice of journal, method of submission, process of review and production of publication are similarly technology-enabled. Much of the publishing activity takes place online with many of the players never meeting face-to-face (Rogerson, 2013). Trusting relationships should exist throughout the publishing activity, but in the online world, this can be challenging. We are fundamentally trusting of each other—such trust is destroyed when an incident occurs that demonstrates untrustworthiness. Trusting relationships in the physical real world rely heavily on non-verbal cues such as body language and tactile interaction, but in the online world, such cues rarely exist (ibid).

It is the virtual space which holds the key to developing trusting relationships in the online world. Academic publishing has moved from what Hine (2000) calls the *cultural artefact view* where technological products are developed and used as tools by those involved in the publishing cycle to the *cultural view* where virtual space enables people to form and reform practices and meaning throughout the publishing cycle, for example, the change of unit of publishing as discussed previously. Humans have become more intimate with technology in this cycle. In this way humans are extended as they are both physical and digital. In this new situation relationship-building takes place through our *digital selves* as it is unlikely that actors in the publishing cycle will ever physically meet.

6.15.4 Editor's Digital Self

For an editor it is essential to embrace the virtual space. It is the editor's *digital self* which helps to build trustworthiness and demonstrate integrity in the eyes of prospective authors. It also increases journal visibility and promotes it beyond the traditional networks of the past. Emerald has been proactive in using digital self. The main elements are Kudos, ORCiD, Twitter and LinkedIn. However, the *virtual anatomy* of my own digital self goes beyond this and includes ResearchGate, Academia.edu, Wikipedia and Google Scholar.

The closest direct links to JICES are through Kudos and ORCiD. Kudos is a web-based service which helps to maximise the visibility and impact of published articles. Emerald encourages editors and authors to use Kudos by linking it to the ScholarOne system. ORCiD provides a unique universal identity for individual academics. Part of the registration process for JICES on ScholarOne for editors, reviews and authors includes the ORCiD identity code. Those without a code can be directed to the ORCiD website to create a code as part of the registration process.

The business social network LinkedIn allows users to create professional profiles, post blogs and interconnect with each other. The *gateway-access approach* to joining LinkedIn helps to build trust among its users. Through LinkedIn blogs I have been able to inform a wide network of people about JICES papers and news as well as current ethical/social issues related to ICT. Conscious of the modern attention-deficit world of the so-called *Generation Y* or *selfie generation* LinkedIn blogs are turned into microblogs through 140-character tweets on Twitter. LinkedIn has led to new contacts and new authors for JICES. LinkedIn is a hybrid element of my virtual anatomy as it helps to develop my reputation as an academic editor and the reputation of JICES as a quality publication.

Two platforms, ResearchGate and Academic.edu are used to make my own academic publications and activities more widely available. Both platforms, whilst having differences, share the common goal of making research accessible for all. As such they are important elements of my virtual anatomy giving me the potential to connect with colleagues, peers and specialists in field relevant to JICES.

The editor of a cross-discipline journal such as JICES must be open to submissions from even the most unlikely of sources. Reaching out to unconventional sources is supported by using some of the generalist channels of virtual space. As such I have used Google Scholar Citations which groups my work together in such a way that access to one of my publications is likely to make others visible to the searcher. The second generalist channel which I use is Wikipedia. It remains the internet's leading encyclopaedia. It may have its critics but when searching for material in virtual space Wikipedia entries will almost certainly be included in the results. Therefore, the final element of my virtual anatomy is my Wikipedia biography.

6.15.5 Future Prospect

In the modern era of academic publishing JICES will flourish because the virtual space has been embraced by both publisher and editor. As its editor the virtual anatomy of my digital self is an interconnected web through which I become credible and creditable and JICES becomes visible and viable.

6.16 Digital Outcasts & COVID-19 [2020]*

This section was first published as: Rogerson, S. (2020) Digital outcasts & COVID19, News & Blogs, *Emerald Publishing, 24 March at www.emeraldpublishing.com/news-and-blogs/digital-outcasts-covid19 Copyright © Simon Rogerson.*

This morning (24 March 2020) is day 17 of my self-isolation thanks to #coronavirus. The world has changed. The social glue has come unstuck and we have turned to technology to allow us to live and keep us connected.

Communication channels keep us informed of the latest developments, advice and restrictions. Social media is keeping our social groups and families together. It is helping to keep spirits up through the likes of dancing policemen in India demonstrating how to wash your hands and dancing nurses in England relieving the tensions with a one-minute disco session. To smile and laugh when we can in these dark times is so important.

But there is a concern. Catchphrases such as global village and digital native have become the reality. From last week we can ask our pseudo friend, Alexa for the coronavirus update from the BBC and "she" will tell us. Local medical surgeries have closed their doors and we now use the NHS App to order prescriptions and seek consultations at a distance with our doctor. Those in isolation turn to online shopping for their groceries, nervously hoping they can find a delivery slot. Virtual social interactions through Facetime, Skype, WhatsApp, Zoom and others are now commonplace. So, what is the concern? It is an often-overlooked digital divide; a divide of two dimensions; young—old and high tech—low tech. Let me deal with these in turn.

Recent statistics puts the worldwide active digital population at 4.54 billion people which is about 59% of the global population. So, 41% of the global population is reliant upon other means to remain informed about the pandemic. This is concerning. However, the most vulnerable to COVID-19 are the elderly and therefore this section of the population deserves attention. Here in the UK 12 million people, about 18% of the population are over 65 years of age. The UK Office of National Statistics has found that 29% of over 65-year-olds have never used the internet. That is 3.48 million people which suggests enormous numbers worldwide are not connected. These vulnerable people do not have the ability to be supported emotionally or medically through the online. They are not digital natives; they are digital outcasts.

The second dimension focuses on access to high technology. Smartphones, virtual assistants, smart TV's and tablets are typical high technology artefacts which provide a multitude of interactions to those who can afford, are digitally literate and are not digitally averse. Low technology offerings such as telephones, radio and tv provide restricted interactions, the latter two only supporting one-to-many broadcasting. The low-tech dependants are at a disadvantage compared with their high-tech counterparts. In times of need they are reliant upon e-buddies within the high-tech population to support them by acting as gatekeepers.

This overlooked digital divide, defined by its two critical dimensions, could have catastrophic consequences in times of emergency such as the COVID-19 pandemic. Understandably governments are focusing on the operationally feasible urgent actions in their attempts to overcome the pandemic. However, the digital divide, discussed in this article, is putting a large proportion of the most vulnerable at even greater risk because actions will not reach them. Once this crisis is over and we reflect on the lessons to be learnt, top of the agenda must be to remove forever this extremely dangerous digital divide. Indeed, digital outcasts must be a thing of the past.

6.17 Is the Digital Divide of the Past, Present or Future? [2020]*

This section was first published as: Rogerson, S. (2020) Is the digital divide of the past, present or future? News & Blogs, Emerald Publishing, 17 March at www.emeraldpublishing.com/news-and-blogs/is-the-digital-divide-of-the-past-present-or-future/. Copyright © Simon Rogerson

Throughout history there have always been social divides predicated upon, for example, poverty, education, gender and status. Such divides have been challenged sometimes by governments, sometimes by radical politicians and sometimes by socially responsible individuals. Often such divides have been diminished by pragmatic people driven by justice and philanthropy. These towering figures who made a difference include Angela Burdett-Coutts, George Cadbury, Andrew Carnegie, George Peabody, Joseph Rowntree, Titus Salt and Louisa Twining.

With increasing technological global dependency, the digital divide has become one of the most significant social divides of our time. Technological advance continues to accelerate and so the digital divide is likely to become more acute with every passing day. In this age of the digital divide there is need for the modern philanthropist who, unhampered by convention, authority and tradition, can help and support the needy. A prime example of the modern philanthropist is Sir Tim Berners-Lee who created the World Wide Web Foundation which focuses on establishing the open Web as a basic right and a public good.

It is twenty-five years since Terry Bynum and I wrote "Cyberspace: the ethical frontier" for the Times Higher Education, formerly The Times Higher Education Supplement. We explained that,

> Information, as the new life-blood of society, empowers those who have it; but it also disenfranchises those who do not. Wealth and power flow to the *information rich*, those who create and use computing technologies successfully. They are primarily well-educated citizens of industrialised nations. The *information poor*—both in industrialised countries and in the third world—are falling further and further behind.

Information often spawns knowledge and understanding. There are three groups of players affected by the information rich—information poor divide; those who create information, those who communicate information and those who consume information. Discrimination, prejudice or bias related to any of these groups is likely to lead to an increase in this aspect of the digital divide. For example, those controlling a communication channel might have a view about who or what might be allowed on the channel. The question is whether that view is morally and socially justifiable. If not, then some information will be unfairly barred from communication thereby curtailing choice in the consuming community.

The impact of such situations is far reaching both in breadth and depth. Here are some examples. Job opportunities which are only advertised online cannot be viewed by those without web access. Personal healthcare monitoring systems with online connections demand a level of computer literacy of the patient, as well as access to the online system. Online training and education courses are only available to those with online access which has sufficient capacity to allow functions such as video streaming and Voice over Internet Protocol (VoIP) as these are an often integral to the course on offer.

However, the digital divide is more than simply information as a product. It is suggested that it is broad in its meaning; referring to unequal access to information and communication technology based on social, economic, cultural and political factors. It is multi-dimensional, encompassing a variety of diverse perspectives and dynamic changes over time as technology evolves. It is commonplace for single dimensions, such as location, gender, age and education, of the digital divide to be considered in isolation. This seems to be problematic as the digital divide occurs through a very complex interrelationship of factors which need to be considered in total. It is this which makes the digital divide one of the greatest challenges of today. Since the advent of accessible online computing, the digital divide existed, it exists today and it will exist tomorrow. It means that almost every aspect of life will be affected, particularly for those who are most vulnerable for whatever reason.

References

Badaracco, J. L., & Useemm J. V. (1997). The Internet, Intel and the vigilante stakeholder. *Business Ethics: A European Review*, *6*(1), 18–29.

Bateman, P. J., Pike, J. C., & Butler, B. S. (2011). To disclose or not: Publicness in social networking sites. *Information Technology & People*, *24*(1), 78–100.

Bhopal, S. S., Bagaria, J., Olabi, B., & Bhopal, R. (2021). Children and young people remain at low risk of COVID-19 mortality. *The Lancet Child & Adolescent Health*, *5*(5), e12–e13.

DeBré, E. (2021, March 25) How a software error made Spain's child COVID-19 mortality rate. *Skyrocket*. Retrieved November 17, 2021, from http://slate.com/technology/2021/03/excel-error-spain-child-covid-death-rate.html.

Ellwood, W. (1996, December 5). Seduced by technology. *New Internationalist*, *286*, 7–10.

Gellman, R. (2001, September 24–26). Public record usage in the United States. In *Proceedings of the 23rd international conference of data protection commissioners*. Commission Nationale de l'Informatique et des Libertés.

Hine, C. (2000). *Virtual ethnography*. Sage.

Johnson, D. G. (1997). Ethics online. *Communications of the ACM*, *40*(1), 60–65.

Light, B., & McGrath, K. (2010). Ethics and social networking sites: A disclosive analysis of Facebook. *Information Technology & People*, *23*(4), 290–311.

Liu, B., & Xiong, L. (Eds.). (1990). *Portraits of ordinary Chinese*. Foreign Language Press.

Louch, M. O., Mainier, M. J., & Frketich, D. D. (2010). An analysis of the ethics of data warehousing in the context of social networking applications and adolescents. *2010 ISECON Proceedings*, *27*(1392).

Moor, J. H. (1998). Reason, relativity, and responsibility in computer ethics. *ACM SIGCAS Computers and Society*, *28*(1), 14–21.

Murgia, M., & Espinoza, J. (2021, November 9). Facebook whistleblower warns UK and EU to do more to control online. *Financial Times*. Retrieved November 17, 2021, from www.ft.com/content/dcc9c9bf-2abe-4167-aaac-efc067d5a359.

Rogerson, S. (2013). The integrity of creating, communicating and consuming information online in the context of Higher Education Institutions. In L. Engwall & P. Scott (Eds.), *Trust in universities*, Wenner-Gren International Series, *86* (pp. 125–136). Portland Press Limited.

Silverthorne, S. (1999). *Online safety rules of the road*. Retrieved October 24, 2021, from http://familypc.zdnet.com/safety/security/feature/21d2/.

Strutin, K. (2011). Social media and the vanishing points of ethical and constitutional boundaries. *Pace Law Review*, *31*(1), 228–290.

Taylor, J. (1995). The networked home: Domestication of information. *RSA Journal*, *143*(5458), 41–53.

Tysome, T. (2004, April 30). "Portal combat," ICT in higher education. *The Times Higher Education Supplement*, *3*, 4.

Witt, C. L. (2009). Social networking: Ethics and etiquette. *Advances in Neonatal Care*, *9*(6), 257–258.

Chapter 7

Synthesis

In July 1999, I was interviewed by Lindsay Nicolle for the BCS and the interview appeared in *The Computer Bulletin* (Nicolle, 1999). I was quoted as saying that professionals involved in the delivery of digital technology,

> have a huge amount of power in modern society, but with that power comes an equally huge responsibility and obligation to do things that are acceptable, and which promote the well-being of every one of us. . . . Economic pressures on profit making companies to seek short-term project gains and consider only the technical and economic feasibility of projects means they compromise on what is socially acceptable—yet long-term benefits of doing otherwise . . . are huge. Considering the social issues, implications and impacts of IT projects can build up enormous consumer trust, brand loyalty and repeat business. That requires organisations to change the way they look at the feasibility of their activities.

Between 1998 and 2010 a series of six surveys was undertaken to capture the opinions of professionals involved in digital technology regarding identifying and dealing with the associated ethical and social issues. Summaries of the six surveys are included in separate sections within this chapter (7.3, 7.5, 7.8, 7.9, 7.10 and 7.13). These, together with a range of other articles, explore the issues raised in the Nicolle interview. In total they form a synthesis of how the digital technology industry's actions and opinions have changed, or otherwise, during 25 years of technological revolution.

The world has now become digitised through, for example, 5G, clouds, AI, algorithms, augmentation, machine autonomy, data analytics, edge computing, and the Internet of Things, these lead to new applications such as smart cities and cryptocurrencies. Thus, there now exists global deep-seated dependency on digital technology. This digitisation of everything requires a greater emphasis on, what we should now call, Digital Ethics which can be defined as integrating digital technology and human values in such a way that digital technology advances human values, rather than doing damage to them. If not, then a very bleak, discriminatory world beckons.

Everyone has moral obligations and responsibilities in ensuring the Digital Age is inclusive and empowering rather than exclusive and constraining. Established rules may offer some guidance as to the correct path, but such rules can easily become the instruments of blatant superficial

DOI: 10.1201/9781003309079-7

compliance which at best is problematic and at worst immoral. It is virtuous action that promotes an ethical digital age (Rogerson, 2020c). The Afterword which follows this chapter offers a way forward in pragmatically addressing the expanding ethical and social issues which surround and, when not acknowledged and addressed, can stifle digital technology potential.

7.1 Computers and Human Values [1996]*

This section was first published as: Rogerson, S. (1996) ETHIcol—Computers and Human Values. IDPM Journal, Vol 6 No 5, pp. 14–15. Copyright © Simon Rogerson

This special edition of ETHIcol reports on a recent public lecture titled "What are Computers doing to Human Values?" given by Professor Terrell Ward Bynum who is the director of the Research Centre on Computing and Society at Southern Connecticut State University, US and has been an influential figure in the field of computer ethics for over ten years.

In his opening remark, Professor Bynum placed IT in a historical context. He argued that IT was in the vanguard of societal change agents saying,

> Powerful technology radically changes the world. Think about what happened when farming was developed, suddenly you could stay in one place, set up trading centres and build large buildings. It revolutionised the world. The industrial revolution is obvious to all as having changed the world radically.

He explained that,

> IT is logically malleable. It can be moulded and adjusted to carry out almost any task. If you can describe a task in such a way that it can be carried out a simple step at a time and you can understand what those steps are, you can actually programme a computer and attach some peripheral devices to carry out that task. Now initially the tasks might be quite mechanical but the task need not be mechanical and one of the nice things about computers is that the equivalent of reasoning, calculating, making analogies and so on can be done by computers as well. So IT being logically malleable is perhaps the most powerful technology ever created. Now if it is the most powerfully created technology then it is going to change the world more fundamentally and given how rapidly its developing, perhaps more quickly than any technology ever devised. IT is the closest thing we have to a universal tool. You can do almost anything with it.

It is Professor Bynum's view that IT will change the world and affect the human values more than any other technology has done in history. IT seeps into our lives unnoticed. Without realising it people become dependent on IT. The reliance on smart cards, credit cards, calculators and programmable domestic appliances are testament to that. He explained that this has happened in two stages.

> The first stage is the introductory stage in which the new technology is invented and then produced and refined. It is very expensive and does not do its assigned role very well but after a while it gets refined and improved and the price comes down. Eventually, it is so inexpensive and proves to be so useful and helpful that it permeates

our society. In the first stage, the technology enables you do the same job perhaps in a more interesting way or safer way or more efficient way.

He continued,

In the second stage, once the technology permeates the society and has its influence everywhere you look, it transforms the meanings of the terms that we use to describe our society and our relationships with each other. It transforms the activities that we engage in as human beings interacting with each other. So basically what we are looking at are social transformations.

This a major point that Professor Bynum elaborated upon using several examples. In discussing work he explained that,

In the Anglo Saxon world, the work ethic is very important. What you do for a living and whether you work hard or you do not work hard are some of the defining characteristics of one's person. Work is being influenced in all sorts of remarkable ways by IT so that you can now do it anywhere and anytime. More jobs are lending themselves to teleworking and that has an incredible impact on a wide variety of circumstances.

For the ordinary person, work is the main way of accumulating wealth but even this human value is changing according to Professor Bynum.

The very concept of who is wealthy is beginning to change. It used to be you were wealthy if you had money, now you are wealthy if you can get goods and services where more and more of those goods and services are available over the information superhighway. Education, banking and medical care are all moving into cyberspace. If it turns out that you are well connected to cyberspace and you can get educational opportunities, you can get job opportunities you can get medical services and you can get entertainment—that is wealth, because there is a whole other part of the population, at the moment about 96%, of the whole world who are not connected and cannot get those benefits. As these benefits move into cyberspace more and more, the gap between the haves and the have-nots gets bigger and bigger and bigger.

Professor Bynum is clearly concerned that IT is having a wide variety of impacts on democracy that are not fully understood by society. He recalled the events of a presidential election explaining that,

The major news networks in the US using computers and exit poles were able to tell at 3.00 in the afternoon Eastern Time that Ronald Reagan had won the presidency of the United States. At 3.00 in the afternoon Eastern Time none of the polls had closed in the Mid-West, the Rocky Mountain states and the West coast. ABC and CBS at about 4.30 in the afternoon, with these polls still open, announced that Ronald Reagan had been elected President. A lot of people who were planning to go to the polls because they were interested in this presidential election did not go, they came home from work and instead of going out to vote, they stayed home and watched the television to hear about how Ronald Reagan had been elected President. Now because they did not go to the polls, it did not affect who was elected President but they did not vote for Governors, they did not vote for Senators, they did not vote for members of the House of Representatives, they did not vote for dog catchers and some Boards of Education. Major studies have indicated that some people won because of that announcement. They would have been voted out, had the people who were planning to go to the polls, gone to the polls. We know what democracy is and we have a pretty good idea of a fair election, is that a fair election? IT made it possible.

All sorts of other interesting things are happening. In the UK and also in the US, there are various experiments where elected officials are wired to their constituents and whenever an important issue comes up to vote on the constituents all send their preferences in to their representatives. Now is not that going to change the very nature of representative democracy? If every major issue becomes a poll that everybody participates in on a weekly or daily basis that is very different from the kind of democracy and representation that we are used to. It is happening but should we let it happen?

Freedom is a human value that is often ignored until it is taken away. Freedom can now be won back using IT. Professor Bynum considered the plight of the severely disabled.

Consider a person who was paralysed from the neck down. Normally that person would require other people to wash, feed and take care of him or her. That person would not be able to hold down a job or function in a normal way. Now take a computer system that is especially designed for such a person to use, for example, if that person can wink an eye it can activate a computer. Then with the right software, you can do word processing, database management you can surf the web, you can do just about anything anybody else can do. Suddenly that person who was totally dependent on other people, could not hold down a job, could no carry on correspondence with others and was not a tax paying money earning worker, could be all of those things. How much does the supporting IT cost? A simple home computer system and a switch that attaches to your eye about £2000. Does our society have any ethical obligation to provide that for that person with a disability?

Arguing philosophically, Professor Bynum showed it was ethical.

Consider the utilitarian argument. The utilitarian says we want to bring about the most happiness and self fulfilment for the most number of people. Consider two societies, one in which you have about 95% of the population who are functioning normally and they are paying taxes and they are working and they are doing all sorts of interesting things with each other and another society where there is only 85% because the 15% who are severely disabled have no access to IT and so we all have to pay for them and take care of them. Which society has, from the utilitarian point of view, the most ethical situation? Obviously more people are fulfilled, more people are happy and thus more people are benefiting in the society where 95% are functioning practically than where only 85% are. So from a utilitarian point of view, ethically the society ought to provide IT to the disabled. From a Kantian point of view, the right thing to do is to maximise one's personal autonomy, ones control over one's life so that one can take responsibility for oneself and decision by decision make a life. Well obviously someone who is empowered by technology is empowered to do that is in a much better position from a Kantian point of view than someone who is not. Indeed, denying someone the ability to be in charge of their own lives and take responsibility for becoming what they have become is grossly unethical from a Kantian point of view.

Governments, organisations, and individual citizens make a grave mistake if they view the computer revolution as technologically neutral. It is fundamentally social and ethical. Each of us has an obligation to consider carefully the impact on human values of each IT application we are planning. The potential and risks alluded to in Professor Bynum's lecture are plain to see by us all.

7.2 Practice and Virtual Behaviour [1997]*

This section was first published as: Rogerson, S. (1997) ETHIcol—Practice and Virtual Behaviour. IMIS Journal, Vol 7 No 4, p. 9. Copyright © Simon Rogerson

The views of the readership are sought on two specific issues, Ethics in Practice and Virtual Behaviour.

7.2.1 Ethical Issues in Practice

The ETHICOMP conference series is a European forum to debate the social and ethical issues related to the application of IT in organisations and society as a whole. The fourth conference in the series will be held in March 1998 at Erasmus University in the Netherlands. The overall theme of the conference is computing and the workplace, the potential tension between, on the one hand, financial goals, politics and personal agendas, and, on the other, social and professional responsibility. It will be an opportunity for practitioners and policy makers to meet and discuss these issues with international scholars working in this field.

A number of topics have been identified as being relevant to IT in the work place. These are summarised as:

- **Organisation structure and the location of work**—As powerful change agents, computer technologies change organisations and social structures. The global community raises many issues relating to ethnic, cultural and economic differences. Computer enabled work.
- **Electronic commerce**
- **Privacy and monitoring**—Issues relating to information held on individuals. What can be or has to be revealed and what safeguards should be in place to ensure privacy. Data Matching. CCTV monitoring. Smart card usage.
- **Value and accuracy of data and information**—Issues of authenticity, fidelity and accuracy and how information can be cultivated for general good. Trusted third parties.
- **Software and data as intellectual property**
- **Security and computer misuse**—Issues of misuse and the impact of misuse of IT.
- **Developing information systems now and in the future**—How to ensure social value issues are properly addressed and how to ensure future advances can be catered for and used in a socially acceptable manner. Responsibility and liability for development and implementation.

■ **Encouraging and enforcing professional standards**—Licensing, codes of conduct and standards of practice.
■ **Socially responsible hardware and software marketing**

Question: Are these issues significant when organisations apply IT to solve business problems and try to improve operational efficiency or service and product delivery? Are there other issues that you would add to the list?

7.2.2 Virtual Behaviour

The second issue where your views are sought relates to working with the internet. According to Deborah Johnson there are three general ethical principles that promote acceptable behaviour in the virtual society:

■ know the rules of the on-line forums being used and adhere to them
■ respect the privacy and property rights of others and if in doubt assume both are expected
■ do not deceive, defame or harass others

The outcome of not subscribing to such principles is likely to result in chaos overwhelming democratic dialogue, absolute freedom overwhelming responsibility and accountability, and emotions triumphing over reason.

Question: Do you agree with these three principles and are there other principles that you would add?

For organisations operating in the virtual society these principles can be expanded into a number of explicit actions such as:

■ Establish an electronic mail policy that forbids forgery of electronic mail messages, tampering with the email of other users, sending harassing, obscene or other threatening email and sending unsolicited junk mail and chain letters.
■ Clarify the responsibilities of those involved in providing information. The publisher is the producer of the on-line information and is, in the main, responsible for that information. The producer therefore must be identifiable at all times. It is arguable that people can only be held responsible for things they can control. Access providers take this line refuting responsibility for the wrong-doings of users, because their contribution as access providers is purely technical.
■ Develop international co-operation that will encourage the creation of common descriptions for internet services, transparency which would benefit users, and respect for trademarks, and increase the possibility of global access to the internet on demand.
■ Encourage the development of legitimate electronic commerce that upholds consumer protection through the use of standard contracts and technical mechanisms. Many issues need to be taken into account including the validity of the electronic signature, the solvency of the purchaser, the legitimacy of the vendor, the security of the transaction and the payment of appropriate sales taxes.
■ Establish a body for handling customer complaints and reviewing internet activity of the organisation. This body would be in contact with public and private international groups that are competent in internet affairs.

■ Train employees in ethical internet practice and promote internet awareness within the wider community.

■ Promote a greater equality of access to the internet through multilingual and multicultural support. This counters the current situation where the internet is an Anglo-Saxon network with 80% of servers being North American and 90% of exchanges taking place in English.

Such actions provide the means for self-regulation that would combat the regulatory flux surrounding the internet and might lead to effective policies regarding transparency, responsibility and respect of appropriate legal frameworks.

Question: Are there other practical actions that you would recommend?

7.3 IS IT Ethical? 1998 Report [1999]*

This section was first published as: Rogerson, S. & Prior, M. (1999) ETHIcol—IS IT Ethical? 1998 ETHICOMP Survey of Professional Practice. IMIS Journal, *Vol 9 No 1, pp. 10–11. Copyright © Simon Rogerson & Mary Prior.*

Companies throughout Europe continue to invest heavily in ICT. It is estimated that Europe's companies spend more than $210 billion annually. Intranets and the internet are seen as the leading corporate technologies. An increasing number of chief information officers now sit on the board of companies. It is clear that the ICT professional is an important person in the workplace and one who has increasing power and influence. It is important that such people strive to work to the highest possible professional standards.

With this in mind the IMIS has long been interested in the ethical and social responsibility issues related to the use of ICT within organisations and by society in general. As part of this interest and activity detailed information has been sought from the membership. A multidisciplined team from the Centre for Computing and Social Responsibility at De Montfort University has investigated the attitudes of information systems professionals to a range of ethical issues. The survey was published in the March 1998 edition of the IMIS Journal. A report of the survey's findings is now available. This edition of ETHIcol highlights some of the key points in the report.

Of the 170 respondents to the survey, 152 are members of the IMIS. The overwhelming majority (88.2%) are male. There is an equal number of responses from those aged between 25–40, and 41–50, with fewer in the over-50 age bracket and fewer still under 25. The majority of respondents are experienced personnel; 67.2% have more than 10 years' experience as an IS professional. The largest single group by job title is "Manager/Director of IS" — 26.6% of respondents

7.3.1 Being Ethical

An encouraging number of respondents—over 90%—either agree or strongly agree that organisations should require all employees to abide by a code of professional ethics, and that employees who violate such a code should be appropriately disciplined. Over 60% of respondents say that their employing organisation has a code of conduct for all employees. However, a much smaller proportion of organisations have a code of conduct for IS employees. Many IS professionals are less than completely satisfied with the existing codes within their organisations.

While it is true that respondents to a questionnaire concerning ethical attitudes are likely to be those interested in ethical behaviour, this nevertheless represents an extraordinary level of support for the implementation of ethical policies at an organisational level, although there could be widely diverging views as to what constitutes "appropriate" penalties among the respondents. A majority also agree or strongly agree that organisations should develop and administer an ethics awareness programme for their employees. The evidence from the survey suggests that it is the younger IS professional who would particularly benefit from such an ethical awareness programme.

7.3.2 Unethical Work

There is a correlation between the type of employing organisation and those people refusing to work on a project considered to be unethical. There was 100% agreement from the self-employed and "others," reflecting perhaps the greater freedom of the self-employed to choose their work. The largest proportion of respondents indicating either disagreement or indifference was among those employed in private enterprise outside the computer industry.

These results may suggest that many of those who care about the type of projects on which they work choose employment in areas such as public service, while many of those who are less bothered find employment in industries in the private sector. Another possible explanation is that employers in private enterprise outside the computer industry give employees less freedom to refuse to work on particular projects.

The other influencing factor is age; a higher proportion of those in the 25–40 age group disagree that they would refuse to work on a project that they considered to be unethical and agree that providing a project makes for an interesting challenge, they do not care about its overall objectives or purpose. As this is a crucial period during which careers are being built, perhaps more employees in this age group find they have to be pragmatic in accepting work than do employees in the relatively well-established older group.

If this is the case, there are some important implications here for organisations and for professional societies. This is because it would seem that an unscrupulous organisation could put pressure on a proportion of employees, and especially some of those within the 25–40 age group, to work on ethically dubious projects or aspects of projects.

7.3.3 Intellectual Property

Concerning the recreation of a product, program or design for a new employer, respondents are clearly split on the issue of the ownership of intellectual property of this nature. Age is the most significant factor connected with the response to this issue; more of those over 50 than under 50 say that employees should not be allowed to recreate a product, program or design for another employer.

Once again, the younger age groups seem to be taking a pragmatic approach. In this instance it may be related to the stage the young IS professionals have reached in their career: they expect to move between employers and acknowledge the likelihood of re-creating intellectual products. Or it may be reflecting a shift in view across different age groups

7.3.4 Using Corporate Computers

Respondents draw a distinction between whether their employer's computing facilities are used for the employee's own profit-making or non-profit activities. A majority agree or strongly agree that it is acceptable to use their employer's computing facilities for non-profit activities if this has no adverse effect on their employer (although more than 15% were indifferent). On the other hand, a much smaller number (13.1%) agree or strongly agree that it is acceptable to use their employer's computing facilities for the employee's own profit-making activities, with less than 6% being indifferent.

7.3.5 *Data Security*

More respondents indicate that their organisation's computer-held data is secure from external than from internal sources. The occupational groups that most frequently reported their organisation's data to be at threat from unauthorised access from internal sources were Database Managers, Technical Services Managers and Network Managers.

It would seem that those most closely involved with the storage, access, and transmission of data have less confidence in security arrangements than IS managers. This is a worrying finding, as the more pessimistic view of security could be the more realistic one, since it is held by those working most closely with the hardware and software systems on which data is held. There is a clear warning here to IS managers to ensure that they acquire a realistic assessment of the effectiveness of their security arrangements.

7.3.6 *Electronic Surveillance*

Over 20% of respondents agree or strongly agree that employers are entitled to use electronic surveillance to monitor employees" performance without their consent. The responses appear correlated with age; the tendency is that the younger the age group, the higher proportion of respondents agree with the statement.

It seems extraordinary to find such a high proportion of respondents seemingly unaware of, or indifferent to, the issues raised by the practice of electronic surveillance by an organisation of their employees without their consent. It would seem that fewer of the younger respondents are aware of the ethical issues raised by this practice. Or it may be that the use of CCTV and other electronic surveillance devices has become so ubiquitous in recent years that its presence is taken for granted by younger people, and they see only the benefits in terms of increased public safety and the contribution made towards the detection of criminal behaviour.

It is especially worrying in the light of the fact that IS professionals are the employees likely to be called upon to install and maintain electronic surveillance systems within an organisation.

7.3.7 *Clients and Users*

A number of statements, worded in slightly different ways, were designed to find out respondents' views of their responsibility to clients and to users. There are some interesting variations in the responses to the different statements.

The vast majority of respondents consider the overall working environment to be part of the IS professional's responsibility. An overwhelming majority of respondents agree or strongly agree that ongoing consultation with representatives of all those affected should occur throughout the information systems development life cycle.

However, nearly half of the respondents agree that when disagreements arise between development personnel and those affected by the system it is the project manager who should have the final say. There is some variation of response according to the respondents' job titles. Interestingly, higher proportions of the project leaders and analysts than of other occupational groups disagree with the statement.

The response to the statement concerning who should have "the final say" when there are disagreements calls into question how the respondents interpret the terms "responsibility" and "consultation." For an ethically acceptable approach to systems development, consultation needs to embrace meaningful involvement of representatives of those affected. Employees should have some input into any redesign of their working environment brought about by the introduction of new information systems.

There is clearly considerable work to be done to persuade IS professionals that all stakeholders in an information systems development project need to be a part of the decision-making process. Those in education and the professional societies have a role to play in encouraging young IS professionals to adopt a more participative approach.

7.3.8 Software Development

A significant minority of respondents agree or strongly agree that it is acceptable for a software contractor, provided with a brief specification, to go ahead and develop the system knowing that in the future re-work under another contract will be essential. Yet to do so would imply a level of deceit to the client.

While most respondents agree that it is not acceptable to cut down on testing effort if a project is significantly behind schedule or over budget, a sizeable minority think that it is acceptable. A higher proportion of the under-25s than of other age groups agree that it is acceptable.

The findings for both the issue of honesty to the client and the issue of the amount of testing effort confirm a trend apparent in the responses to some of the other statements, for a higher proportion of younger respondents to adopt a less ethically acceptable approach.

7.3.9 Recommendations

The findings of the survey point to a high level of ethical awareness among those information systems professionals who responded. Nevertheless, there are a number of areas where more guidance and support for individuals is desirable, to encourage consistently responsible behaviour. The survey includes a number of recommendations for organisations. These are intended to promote more socially responsible practices within the information systems community.

It is recommended that organisations should:

- promote awareness among all employees of:
  - ethical issues
  - the organisation's Code of Conduct (if there is one)
  - how the organisation's Code of Conduct may be applied to guide ethical decision-making
- provide a working environment that encourages ethical practices, supporting employees in resisting the temptation to allow commercial pressures to lead to ethically dubious practices—instead, promoting their ethical stance to their commercial advantage
- consider the promotion of ethical awareness among younger IS professionals, including the use of more experienced personnel as mentors
- establish "whistleblowing" procedures to encourage employees who become aware of unethical practices within the organisation to come forward
- introduce a clear policy concerning the use of computing resources by employees for their own activities, and consider allowing the use for selected non-profit-making activities as a contribution to the local community or as a legitimate perk for employees

- review the security of computer-held data with attention to both technical aspects and management aspects affecting potential threats, especially from internal sources
- consider and clarify their policy concerning the re-creation of intellectual property such as a product, program or design by employees when they move to another employer
- seriously consider adopting a Code of Conduct for all employees; and if adopted, include in the Code of Conduct provision for full consultation with employees before initiatives such as electronic surveillance are introduced, and ensure that safeguards for employees are put in place

7.4 The Impact of Change [2001]*

This section was first published as: Rogerson, S. (2001) ETHIcol—The Impact of Change. IMIS Journal, Vol 11 No 4, pp. 25–26. Copyright © Simon Rogerson

This edition of ETHIcol reports on some of the issues raised at the recent ETHICOMP 2001 conference held at the Technical University of Gdansk, Poland. The conference was attended by around 100 delegates from 17 countries. There were 80 papers presented at the conference. ETHICOMP 2001 was supported by IMIS and The British Council.

Systems based on computing technology are powerful change agents in everyday society. We need to consider not only the technological and economic issues but also the ethical and social issues when developing and implementing such systems. It is against this backdrop that the ETHICOMP conference series operates. The overall theme for ETHICOMP 2001 was "Systems of the Information Society." The conference focussed on the ethical and social impacts of these systems on society, organisations and individuals. This was done from four perspectives:

- **Software engineering:** systems development and the relationship between quality, risk and ethics.
- **Teaching and learning:** ethics education for computing students who are the professionals of tomorrow.
- **Virtual communities:** social norms and tendencies of the internet and the impact on families, friends, strangers, traders and consumers.
- **Citizens**: the ethical impacts of computers systems on citizens in a variety of contexts such as the office, the factory, the school, and public areas.

In his keynote address, Don Gotterbarn drew these perspectives together. He explained that sometimes,

> Software is considered to have failed even though it was produced on schedule within budget and met the customer's specified software requirements. Software has been developed which, although meeting stated requirements, has significant negative social and ethical impacts. The Aegis radar system, for example, met all requirements that the developer and the customer had set for it. The system designer's did not take into account the users of the software nor the conditions in which it would be used. The system was a success in terms of budget, schedule, and requirements satisfaction, even so, the user interface to the system was a primary factor in the Vincennes shooting down an Iranian commercial airliner killing 263 innocent people.

He suggested that,

> There are two factors that contribute to these professional and ethical failures. There is significant evidence that many of these failures are caused by limiting the consideration of relevant system stakeholders to just the software developer and the customer. This limited scope of consideration leads to developing systems that have surprising negative effects because the needs of relevant system stakeholders were not considered. In the case of the Aegis radar system the messages were not clear to the users of the system operating in a hostile environment. These types of failures also arise from the developer limiting the scope of software risk analysis just to technical and cost issues. A complete software development process requires the identification of all relevant stakeholders and broadening the risk analysis to address social, political, and ethical issues. Software development lifecycle methods include a risk analysis process but with current methods limit the types of risks considered. The risk analysis is primarily instrumental-addressing corporate bottom lines. Software projects have ethical dimensions that need to be identified before and during the development process.

Clearly there is a need to provide processes and tools to address such issues in the development cycle but that is only half the solution. There has to be a willingness to use such facilities in order to address these broader societal issues. This requires our future professionals to be educated in such matters. The introduction of civics into schools" curriculum would provide the necessary foundation on which to build appropriate elements into higher education programmes designed to produce the computer professionals of the future. In their paper, Eva Turner and Paula Roberts concluded that, "ethics should be part of the training of both the computing professional and the general user, if the working culture of the computer industry, and society's passive acceptance of computers is to be changed." They argued that the topics covered "must include such important issues as gender, race, disability and culture." They suggested that this is an ongoing process of education and that,

> Developing IT technologies will present professional and user alike with the need to continuously adapt to new ways of working and living with computers. Alongside this lifelong flexibility must be the ethical skills to critically evaluate the newly developed technologies on societies.

The stark reality of technological influences on society was illustrated by a case study presented by Helen Richardson and Kate Richardson. Their paper catalogued "the rise and rise of call centre in the North West of England, UK and their use of [Customer Relationship Management] CRM systems. CRM implies new technologies and new ways of working." They discovered that there were "some inherent contradictions in terms of privacy, communication richness, management methods and computer ethics" in using CRM systems. They pointed out that, "Call centres today are often viewed by some as offering satisfying employment of intrinsic value, for others, they are the "new sweatshops of the 21st century."" The clear message from this case study is just how vulnerable certain sections of society are to technologically supported exploitation. In his closing remarks to the delegates of the conference John Weckert explained that situations such as this called for professionals to properly consider issues of equality and justice, freedom of speech and informed consent. They need to subscribe to the notion of responsibility and duty of care when developing systems for society.

The message is simple. An inclusive Information Society needs an infrastructure built upon systems which are in harmony with all citizens.

7.5 IS IT Ethical? 2000 Report [2001]*

This section was first published as: Rogerson, S. & Prior, M. (2001) ETHIcol—IS IT Ethical? 2000 ETHICOMP Survey of Professional Practice. IMIS Journal, Vol 11 No 6, pp. 27–28. Copyright © Simon Rogerson & Mary Prior.

Should IS professionals actively consult all stakeholders in an information systems development project or merely keep them informed? This is one of the questions debated at the recent launch of the IMIS-sponsored report, *IS IT Ethical?* The report is the second in a longitudinal study of the ethical attitudes of IS professionals being conducted on behalf of the Institute by the Centre for Computing and Social Responsibility at De Montfort University.

Both of the surveys carried out to date (in 1998 and 2000) found that the vast majority of respondents agreed on fundamental issues such as the importance of ethical considerations to organisations and to themselves. Nearly all of them, for example, in the 2000 survey agreed that organisations should require IS/IT employees to abide by a code of professional ethics (for the IMIS Code of Conduct go to the IMIS website). Over 80% would refuse to work on a project they considered to be unethical—despite the fact that some 40% said they have no choice about the projects they work on. Most (though not all) said that the unauthorised copying of software was not acceptable practice.

Of course, there may be a distinction between what people say in surveys, even anonymous ones, and what they do in practice. Which perhaps makes some of the other responses even more interesting. For example, nearly 20% agreed it was acceptable, "*for a software contractor, provided with a brief specification, to go ahead and develop the system knowing that in the future re-work under another contract will be essential*"—implying support for a level of deceit to the client. Likewise, although the majority of respondents agreed it was *not* acceptable to cut down on testing effort when over budget or behind schedule, a sizeable minority of 16.6% thought it was acceptable, with as many as 10% indifferent. A similar proportion of respondents considered it acceptable to access data they are not authorised to see by using another employee's access code with their permission. Given the ubiquitous nature of computer systems and the quantity of personal data they now hold, these figures are somewhat worrying.

7.5.1 Profile of Respondents

It is worth noting the profile of the survey respondents. Over 60% are over 40 years of age, 65% have more than 10 years' experience as an IS professional, over 80% are male and the largest single occupation is "Manager/Director of IS." Both the private and public sectors are represented along with a variety of size of organisation. The majority are based in the UK, although responses were received from a number of other countries.

No clear relationship can be established between the respondents" age, experience or occupation and their response to particular issues; the older and more experienced ones were just as likely as their younger colleagues in the 2000 survey to make what the authors" would see as a "less ethical" response.

7.5.2 Employee Surveillance

Another contentious area is that of employee surveillance. Concerned to find in the 1998 survey that as many as 22% of respondents agreed that employers were entitled to monitor employees" performance without their consent, the 2000 survey asked about this issue in more detail. The responses indicate that employees" consent and knowledge are seen as crucial by the majority of respondents. While a majority of nearly 80% agree that employers are entitled to electronically monitor their employees with both their knowledge and consent, the figure drops to 13.9% with neither. This figure, although it represents only a minority of respondents, is still a matter for concern. It is IS professionals who may well be called upon to develop, install and maintain some types of electronic surveillance software and equipment. Quite apart from the ethical issues involved such practice could now fall foul of legal requirements under the Data Protection and Human Rights Acts.

Evidence of an attitude showing greater concern for privacy is found in the 17.5% of respondents who find it unacceptable to electronically monitor employees even with both their knowledge and consent. These findings suggest that there is a need for a much wider debate about the use of electronic surveillance methods in the workplace particularly because, with the evolution of more sophisticated means of workplace surveillance, it is no longer only employees carrying out routine, repetitive tasks who are monitored but also knowledge workers such as IS professionals themselves.

7.5.3 Licensing

Another area where the survey has highlighted the potential for further debate is that of the licensing of IS professionals. Over 40% of the UK respondents agreed with the idea of licensing, 31.2% disagreed while 25% were indifferent. The difficulties of licensing with such a diverse group as IS/IT workers are well known; however, the concept of distinguishing between different levels of worker and licensing those responsible for "signing off" specifications or (parts of) working systems was aired at the report launch. There is clearly a need for ongoing debate on this difficult but important issue.

7.5.4 Consult All Stakeholders

So, should IS professionals actively consult all stakeholders in an information systems development project or merely keep them informed? Respondents to the 2000 survey were fairly evenly split between those who agreed, and those who disagreed, with the statement, *"Consultation with all stakeholders in an information systems development project is not always possible; to keep stakeholders informed is sufficient."* The report's authors maintain that for an ethically acceptable approach to systems development consultation needs to embrace meaningful involvement of representatives of all those affected. Employees should have some input into any redesign of their working environment brought about by the introduction of new information systems. Many IS professionals, however, regard this as idealistic when faced with commercial and economic pressures. Until they can be persuaded otherwise, we are likely to continue to find high-profile information system failures headlined in the media.

7.6 IS Staff and Privacy and Data Protection [2002]*

*This section was first published as: Rogerson, S. & Howley, R. (2002) ETHIcol—IS staff and privacy and data protection. IMIS Journal, Vol 12 No 6, pp. 29–30. Copyright © Simon Rogerson & Richard Howley.

Increasing attention is being given to the contribution that information systems (IS) staff can make to the implementation of the 1998 Data Protection Act and the 2001 Freedom of Information Act in the UK. A focus of this attention has been on the contributions that can be made in the systems design process and in the application of privacy enabling technologies (PETs).

Recent research by the Centre for Computing and Social Responsibility (CCSR) found considerable evidence that IS staff are promoted as key providers of privacy and data protection (PDP) both within organisations and within information systems areas. Literature from Europe, the US and the UK highlights the importance of designing systems for PDP compliance, encouraging the application of PETs and relating data management strategies to the provision for PDP. Whilst witnessing the emergence of

PDP legislation CCSR wanted to know the extent to which this responsibility, that was increasingly being articulated, was known about and accepted by IS staff in UK based organisations. If IS staff are becoming increasingly responsible for PDP in organisations the extent to which they are aware of their perceived contribution and the degree to which they support it will be critical to its realisation. The research addresses these issues by focusing on three key questions.

7.6.1. Are IS staff aware that the responsibility for PDP is increasingly being devolved to them and is it perceived by them as legitimate extension to their role?

Ninety five percent of respondents regard involvement in PDP as a legitimate activity for IS staff and 85% believe that it is an increasingly important part of their work. The acceptance of PDP as a legitimate part of their work is further evidenced by the nature of the involvement reported. Staff report considerable involvement in the area of PDP management and strategy. In more than 54% of organisations represented in the sample IS staff were "prominent in formulating and implementing PDP polices." In 30% of organisations IS staff are "primarily responsible for PDP within [their] organisations." This involvement evidences a considerable acceptance by IS professionals of their role in the provision for PDP. It also suggests that their involvement is much wider than that proposed in the literature.

7.6.2. Which IS roles do IS staff consider to have the greatest contribution to make to the provision for PDP?

Systems design and the application of PETs are widely reported in the literature as areas in which IS staff can contribute to PDP provision. The views of IS staff were sought with regard to which staff roles offer the greatest opportunity to contribute to the provision for PDP. The roles and number of times they were identified are given in the Table 7.1.

Table 7.1 IS roles in the provision for PDP

IS Role	Number of respondents identifying role
IT/MIS/Systems Manager	23
Systems Analyst	15
Systems Developers	9
Database Administrator	9
Network Manager	6
Systems Administrators	6
Project Manager	7
Programmers	3
IT Security Personnel	5
Systems Designer	3
Support and Training	2

The single most important finding was the extent to which staff identified the role of management as critical in the provision for PDP. Indeed, some respondents felt so strongly about this, they annotated their response to add greater emphasis to their answers. The role of the systems analyst (including requirements analysis) is also prominent in findings and it is interesting to relate this to the prominence of "systems design" as opposed to "systems analysis" in the literature. We should not be seduced into thinking that we can "design for compliance" if we are failing to manage the capture and realisation of requirements that are in themselves PDP compliant.

7.6.3. What stages in a systems development process do IS staff feel offer the greatest potential for embedding PDP compliance in information systems?

In the introduction it was reported that systems design and the application of PETs are frequently identified in the literature as key areas in which IS staff can contribute to PDP within organisations and their systems. The views of IS staff were sought with regard to the stages of the systems development process that they feel offer the greatest opportunities for PDP leverage. Table 7.2 presents the findings.

The role of management, which was identified so frequently in response to the previous question, may support the further identification of "project planning" and "embed in the whole process" in the responses to this question. Systems (and or requirements) analysis is prominent in these findings offering further support for the roles identified earlier. This is interesting in that this is a stage that may be presumed to occur before design, even in "rapid" and or "iterative" development environments, and as such the prominence of design in the compliance strategy may be inadequate without an equal emphasis on the analysis process. Whilst design is important, IS staff feel that systems/requirements analysis and project management are equally important, and they should not therefore be neglected by a focus on systems design in isolation of requirements analysis and overall project management.

Table 7.2: Stages in the systems development lifecycle that offer opportunities for PDP enhancements

Stage	Number of respondents identifying stage
Project Initiation and Planning	7
Feasibility Study	3
Systems Analysis	18
Systems Design	23
Coding/Programming	4
Testing	8
Implementation	6
Training users	3
Embed in the whole process	11

7.6.4 Conclusion

There is considerable support by IS staff for their involvement in the provision for PDP. Indeed, they already occupy important strategic PDP positions in many organisations. IS staff are able to identify the stages in a systems development process which offer potential for PDP enhancements and the staff that have the greatest contribution to make. However, there is evidence that certain issues need to be addressed if we are to benefit fully from the contribution of IS staff. Levels of PDP awareness amongst IS staff is felt to be low; more than 50% of respondents felt that IS staff awareness of the 1998 Data Protection Act is not high. IS staff feel that the level of training in PDP issues offered by organisations was low; only 3% of respondents felt that organisations are providing suitable training in PDP issues for their employees. Regarding support offered by professional bodies only 29% of respondents felt that they provide appropriate guidance for members. Could such bodies do more?

There is a management challenge that extends beyond the role of IS staff; management have a responsibility to create and maintain a PDP culture within organisations that positively impacts on all stages of IS development, the operation of information systems and all staff. IS staff alone cannot bring about PDP compliance through some form of technical wizardry, and nor can management. The provision for PDP in organisations and within information systems has to be the result of a holistic cultural and structural commitment to PDP that is bought about and maintained by senior management within organisations. No one group of staff can affect PDP alone—it has to be an organisational wide commitment and be embodied in the very core of the organisation; this is the management challenge that must be addressed.

7.7 Human Rights in the Electronic Age [2003]*

This section was first published as: Rogerson, S. & Beckett, R. (2003) ETHIcol—Human Rights in the Electronic Age. IMIS Journal, Vol 13 No 3, pp. 27–28. Copyright © Simon Rogerson & Robert Beckett.

We are living in an electronic age that raises new challenges to our perception and promotion of human rights. Online communication, media and technology are changing the way we learn, socialise, work, interact, trade, organise, and so on. We are only at the verge of knowing what implications this will have for fundamental human rights and the way we live together in the future. There are many complex questions that need to be answered. Here are just a few.

Question: *How do we balance public safety and individual wellbeing in the digital universe?*

In the global village, each individual has rights to freedom of liberty, expression and conscience guaranteed by the UN Declaration of Human Rights (1948). However, these rights are routinely and continually suspended by national governments and exclusions negotiated by private interest groups.

Question: *How do we establish rights that are based on universal principles and are upheld by governments and private enterprise across new electronic borders?*

Traditionally a key mechanism for upholding rights has been the judicial system, however, the speed and boundary hopping power of electronic systems means that legal systems represent a localised, often long term solution for the worst kind of abuses. The legal timeframe is always much longer than the technological one. This is not an adequate solution. Clearly a new class of rights and mechanisms for their protection are necessary, catering for a new electronically-enabled citizen.

7.7.1 New Form

Consider for one moment this new form of communication. Traditional broadcasting is a one way, one-to-many activity which can be subjected to a range of controls and constraints. Whereas digital media is a two way many-to-many activity in which senders and receivers interchange roles. By its very nature it is uncontrollable in many respects.

Question: *Can we apply existing standards and social norms to this new world?*

7.7.2 Information Moguls

Information providers have taken on a significantly important role in the digital universe. These providers can be uncontrollable by governments because of their global reach and operation. Such providers have great power and influence through deciding information content and format.

Question: *Is society's over dependence on such information moguls acceptable?*

7.7.3 Digital Divide

Information is the new life blood of society and its organisations, and our dependence grows daily with the advance of computer technology and its global application. Information empowers those who have it; but it also disenfranchises those who do not. Wealth and power flow to the "information rich," those who create and use computing technologies successfully. They are primarily well-educated citizens of industrialised nations. The "information poor"—both in industrialised countries and in the developing world—are falling further and further behind.

Question: *Is it right that we tend to promote this divide by our inaction in many cases and by our unwitting activity in others?*

7.7.4 Communication

Let us turn to communication. Article 19 of Declaration of Human Rights states that, "Everyone has the right to freedom of opinion and expression; this right includes freedom to hold opinions without interference and to seek, receive, and impart information and ideas through any media and regardless of frontiers." Whilst Article 5 states, "No one shall be subjected . . . to cruel, inhuman or degrading treatment."

Question: *How do we balance freedom of speech on the internet and use of the internet for purpose of incitement of racial hatred, racist propaganda, homophobia, xenophobia and related intolerance?*

In the name of protecting some civil or human rights, some governments appear to reduce the value and freedom of other rights. For example, "Human Rights Watch reported that Chinese authorities have issued more than sixty sets of regulations to govern internet content since the government began permitting commercial internet accounts in 1995 . . . describes recent Chinese efforts to police internet cafes . . . cases of several people put on trial or sentenced to prison for downloading or posting politically sensitive material on the web." Human Rights News, September 2001.

Blocking, filtering, and labelling techniques can restrict freedom of expression and limit access to information. Government-mandated use of such systems violates rights regarding freedom of speech. Global rating or labelling systems reduce significantly the free flow of information. Efforts to force all internet speech to be labelled or rated according to a single classification system distort the fundamental cultural diversity of the internet and potentially lead to domination of one set of political or moral viewpoints.

Question: *Is it right to employ such techniques in a universal manner?*

Self-regulatory controls over internet content have been promoted by some as an alternative to government regulation. In effect Internet Service Providers, as a group, are being asked to regulate the speech of their customers. The role of an ISP is crucial for access to the internet and because of the crucial role that they play; ISPs have been targeted by law enforcement agencies in many countries to act as content censors. ISPs should provide law enforcement reasonable assistance in investigating criminal activity, confusing the role of private companies and police authorities risks substantial violation of individual rights (The Internet and Human Rights: An Overview, The Center For Democracy & Technology, 5 January 2000).

Question: *Is it right that ISPs become de facto censors?*

"Both self-regulatory and technical solutions are two-edged swords, which can be used to defend information freedom—or to curb it. The government of Singapore, for instance, uses filtering systems to police and censor the internet. Few would argue with parents" right to protect children from illegal and harmful content at home and at school. But what of those prohibitive parents who regard information about worldviews which challenge their own—evolution or atheism, for example—as harmful? ICT thus raises very complex and interesting questions about the application of fundamental freedoms, which require much further debate." A European Way for the Information Society, Report of the Information Society Forum, 2000.

7.7.5 Conclusion

A new Information Society should be based on fundamental assumptions about information and human information rights, access, privacy, self-determination, personal control and not privilege, power, special interest. According to The Centre for Democracy & Technology in their report "The Internet and Human Rights: An Overview," 5 January 2000 governments should subscribe to certain core principles regarding electronic rights.

- Prohibiting prior censorship of on-line communication.
- Requiring that laws restricting the content of on-line speech distinguish between the liability of content providers and the liability of data carriers.
- Insisting that on-line free expression not be restricted by indirect means such as excessively restrictive governmental or private controls over computer hardware or software, the telecommunications infrastructure, or other essential components of the internet.
- Including in the global internet development process citizens from countries and regions that are currently unstable economically, have insufficient infrastructure, or lack sophisticated technology.

- To ensure that internet services are designed to ensure universal access, governments should provide full disclosure of information infrastructure development plans and encourage democratic participation in all aspects of the development process.
- They should also advocate widespread use of the internet and strive to provide adequate training.
- In addition, governments should urge citizens to take an active role in public affairs by providing access to government information online.

■ Prohibiting discrimination on the basis of race, colour, sex, language, religion, political or other opinion, national or social origin, property, birth or other status.

■ Ensuring that personal information generated on the internet for one purpose is not used for an unrelated purpose or disclosed without the person's informed consent and enabling individuals to review personal information on the internet and to correct inaccurate information.

■ Allowing on-line users to encrypt their communications and information without restriction.

■ In order to guarantee the privacy of on-line communication, governments should put in place enforceable legal protections against unauthorized scrutiny and use by private or public entities of personal information on the internet.

- In addition, governments should conduct investigations on the internet pursuant only to lawful authority and subject to independent judicial review.

We need to be proactive about promoting and defending our electronic rights. Abiding by such principles would be a good start. Quite simply, communication without moral application is at best a wasted opportunity and at worst a dangerous threat to society and the rights of its citizens.

7.8 IS IT Ethical? 2002 Report [2003]*

This section was first published as: Rogerson, S. & Prior, M. (2003) ETHIcol—IS IT Ethical? 2002 ETHICOMP Survey of Professional Practice. IMIS Journal, Vol 13 No 6, pp. 29–30. Copyright © Simon Rogerson & Mary Prior.

In September, the third bi-annual study into the attitudes of Information Systems (IS) professionals, "IS IT Ethical? 2002 ETHICOMP Survey of Professional Practice" was launched. Here are some of the key findings.

7.8.1 Respondents

Of the respondents to the 2002 survey just 35% are from the U.K. and 54% from African countries, most of these being from Zambia (28%) and Kenya (15%). The number of respondents aged under 25 has more than doubled to 30% from the previous survey while the proportion of those aged 41–50 has continued to drop with each survey. The vast majority of student members are following part-time study for IMIS qualification whilst in full-time employment within the IT sector. There is still a disappointing level of response from women members.

7.8.2 Ethical Considerations

Three-quarters of respondents agree or strongly agree that they would refuse to work on a project that they considered to be unethical. A larger difference is found in the response to the statement, "*Providing a systems development project provides me with an interesting challenge, I do not care about its overall objectives or purpose."* In 2002 78% of respondents disagree or strongly disagree with this statement as compared to 91% in 2000 and 84% in 1998. As many as 11% are indifferent in the latest survey. Further analysis reveals that of the 14% of respondents who disagree or strongly disagree that they would refuse to work on a project that they considered to be unethical, and the 11% who are indifferent to the statement, *"Providing a systems development project provides me with an interesting challenge, I do not care about its overall objectives or purpose,"* three-quarters are under 40. This appears to support the suspicion raised by the first survey, that there is a relationship between age and the responses to these issues. This significant minority of this indifferent to ethical behaviour is concerning.

7.8.3 Intellectual Property

It would appear that more experienced employees and those working within the public and computer industry sectors are more likely to think that employees should not be allowed to recreate a product, program or design for another organisation on moving from one employer to another.

7.8.4 Using Employer's Facilities

The data suggests that geography is most likely to be having an effect on responses regarding using of employer's facilities, or being in a younger age group and/or being a student in certain countries has an effect on attitudes to use of employer's facilities, but further evidence is required to support this possibility. Use of the internet and use of email are the areas reported as most likely to be subject to a formal policy. There has been an increase in the proportion of respondents who say that their organisation has either a formal or informal policy for the use of software and of printers and peripherals.

7.8.5 Privacy and Security

The vast majority of our respondents do appear to appreciate the boundaries of data access. However, there is a minority who may be prepared to use another's access code with their permission to access data they are not authorised to see. The implication for organisations is that they need to ensure that their privacy and security policies are clear, communicated to all employees, and that employees" awareness and deployment of them are continually reviewed. A higher proportion of respondents say that their organisation's computer-held data is more secure from *external* than from *internal* sources. In the survey 5 out of 7 Network Managers disagreed that data was safe from internal sources, 3 out of 7 disagreed that it was safe from external sources.

7.8.6 Users and Clients

A majority of respondents consider the overall working environment to be part of the IS professional's responsibility. It is a cause for concern that there continues to be a group of respondents who endorse a business practice that lacks openness with the IT services customer. Equally worrying is the finding that there continues to be a minority of respondents who find it acceptable to cut down on testing effort if a project is significantly behind schedule or over budget:

7.8.7 Globalisation

In developed countries, 69% are in favour of and 7% opposed to producing software more quickly compared with 17% and 45% respectively for other countries. For the 25–40 group 11% more are in favour of cheaper software than being in favour of software produced more quickly. The over 50 age group hold the most extreme view with 70% being in favour of cheaper software and 75% being in favour of software produced more quickly. Such views, as mentioned previously in the 2000 survey, may well exacerbate problems of wealth distribution unless there is a reinvestment of profit in the economically poor regions.

7.8.8 Recommendations

It is important for organisations to take account of these trends and look carefully at the following recommendations:

- Seriously consider adopting a Code of Conduct for all employees
- Promote awareness among all employees of:
  - ethical issues
  - the organisation's Code of Conduct
  - how the organisation's Code of Conduct may be applied to guide ethical decision-making
- Establish "whistleblowing" procedures to encourage employees who become aware of unethical practices within the organisation to come forward
- Introduce a clear policy concerning the use of computing resources by employees for their own activities, and consider allowing the use for selected non-profit-making activities as a contribution to the local community or as a legitimate perk for employees
- Establish clear guidelines for the introduction and operation of any electronic surveillance process, including email and internet usage monitoring, ensuring that all employees are fully consulted and that their rights to privacy in the workplace are respected
- Review on a regular basis the security of computer-held data with attention to both technical aspects and management aspects affecting potential threats
- Consider and clarify their policy concerning the re-creation of intellectual property such as a product, program or design by employees when they move to another employer
- Ensure that all policies are clearly communicated to employees and are deployed throughout the organisation
- Promote an approach to systems development that encourages genuine stakeholder involvement in decision-making
- Promote a high level of data protection awareness among staff and review the means by which compliance with data privacy and data protection requirements are assured
- Provide a working environment that encourages ethical practices, supporting employees in resisting the temptation to allow commercial pressures to lead to ethically dubious practices—instead, promoting their ethical stance to their commercial advantage

7.9 IS IT Ethical? 2004 Report [2005]*

This section was first published as: Rogerson, S. & Prior, M. (2005) ETHIcol—IS IT Ethical? 2004 ETHICOMP Survey of Professional Practice. IMIS Journal, Vol 15 No 5, pp. 31–32. Copyright © Simon Rogerson & Mary Prior.

In September the fourth bi-annual study into the attitudes of Information Systems (IS) professionals, "IS IT Ethical? 2002 ETHICOMP Survey of Professional Practice" was launched at ETHICOMP 2005 in Sweden. A copy of the survey was sent to all IMIS members and was also made available on the IMIS website. The country profile of respondents reflects the Institution's strong presence in African countries such as Zambia and Kenya and its high proportion of student members.

The survey findings suggest a high level of ethical awareness among the IMIS members who responded, who overwhelmingly agree that ethical issues are important to organisations and to themselves as individuals, that they would not make unauthorised copies of software to use at work nor at home, and who would not use other employees" access codes without their permission to view data they are not authorised to see. However, the findings also highlight areas where more guidance and support for individuals is desirable to encourage consistently responsible behaviour.

Among the findings that will be of interest to employers is that respondents consider their organisation's computer-held data to be more secure from *external* than from *internal* sources. Taken with the finding that some respondents consider it acceptable to use other employees" access codes *with* their permission to access data they are not authorised to see, this provides a warning to organisations to review both the technical aspects of the security of their systems and also the human resource management issues.

Employers may also be interested to find that over the four surveys, a large proportion of respondents continue to consider it acceptable to use their organisation's computing facilities for their own *non-profit-making* activities providing this has "no adverse effect" on their employer. This is the case despite the widespread existence of organisational policies covering facilities" use. Organisations may wish to ensure that employees are not only informed about policies but that they understand their relevance; for example, what are the possible "adverse effects" of unauthorised use. A regular review of policies and frequent reminders to employees may also be required.

A cause for concern is the finding that the minority of respondents who find it acceptable to cut down on testing effort if a project is significantly behind schedule or over budget is continuing to grow with each survey, reaching 22% in 2004. It is somewhat alarming to consider the consequences on software quality and the potential impact of such an attitude. Another worrying result is that the proportion of respondents who consider it acceptable for employers to use electronic surveillance to monitor employees" performance *without* either their knowledge or consent has risen to 20%.

The report contains recommendations for organisations, for professional societies and for educators that are intended to promote more socially responsible practices within the IS community.

It is recommended that organisations should:

- seriously consider adopting a Code of Conduct for all employees
- increase efforts to promote awareness among all employees of:
  - ethical issues
  - the organisation's Code of Conduct
  - how the organisation's Code of Conduct may be applied to guide ethical decision-making
- establish "whistleblowing" procedures to encourage employees who become aware of unethical practices within the organisation to come forward
- introduce a clear policy concerning the use of computing resources by employees for their own activities, and consider allowing the use for selected non-profit-making activities as a contribution to the local community or as a legitimate perk for employees
- establish clear guidelines for the introduction and operation of any electronic surveillance process, including email and internet usage monitoring, ensuring that all employees are fully consulted and that their rights to privacy in the workplace are respected

- review on a regular basis the security of computer-held data with attention to both technical aspects and management aspects affecting potential threats
- consider and clarify their policy concerning the re-creation of intellectual property such as a product, program or design by employees when they move to another employer
- ensure that all policies are clearly communicated to employees and are deployed throughout the organisation
- promote an approach to systems development that encourages genuine stakeholder involvement in decision-making
- improve the promotion of data protection awareness among staff and review the means by which compliance with data privacy and data protection requirements are assured
- make greater efforts to provide a working environment that encourages ethical practices, supporting employees in resisting the temptation to allow commercial pressures to lead to ethically dubious practices—instead, promoting their ethical stance to their commercial advantage
- consider whether they have adequate guidelines stating what information should be available on their intranets
- ensure that information about individuals is not made available on their intranet without the specific consent of those involved

It is recommended that professional societies representing the IS profession should:

- ensure that their Code of Conduct remains up-to-date and relevant to the profession, increasing efforts to promote awareness of the Code among members and providing guidance how it can be applied in practice
- provide a greater degree of particular support for their younger members, helping them to acquire greater awareness of the ethical issues they will encounter throughout their careers
- promote debate of the continuing applicability of legislation such as the software licensing laws, in the light of current developments, opinion, and practice
- promote debate concerning the desirability of licensing for information systems professionals
- promote members" awareness of the role that IS staff play in the designing of data privacy and data protection compliance into information systems

It is recommended that the academic community responsible for the education of future IS professionals and for research into IS-related issues should:

- address ethical issues more extensively in their curriculum, to raise the awareness of young, aspiring professionals concerning all of the issues covered in this survey
- include in their research agenda the issues identified by this survey as requiring further investigation, for example:
  - the effects of factors such as age, geography, and culture on attitudes to issues such as intellectual property, the use of employer's computing facilities, software and systems testing and the licensing of IS professionals
  - the factors affecting the IS professional's ability to choose and choice of what projects they work on
  - the effect of workplace monitoring on employees and guidelines for the ethical use of employee surveillance

- the implications of the globalisation of the IS profession and recommendations for socially responsible practice in this area
- the applicability of intranets for providing different types of information

7.10 IS IT Ethical? 2006 Report [2008]*

This section was first published as: Rogerson, S. & Prior, M. (2008) ETHIcol—IS IT Ethical? 2006 ETHICOMP Survey of Professional Practice. IMIS Journal, Vol 18 No 1, pp. 26–27. Copyright © Simon Rogerson & Mary Prior.

The fifth bi-annual study into the attitudes of Information Systems (IS) professionals, "IS IT Ethical? 2006 ETHICOMP Survey of Professional Practice" is now published. This survey was somewhat different from that of the previous ones, with the consequence that the profile of respondents is also dissimilar. There is a much smaller number of responses from industrial practitioners who are members of IMIS this time. The majority of respondents are young people studying for a computing degree in the UK. Other respondents include a group of experienced academics attending the ETHICOMP 2007 conference and, for the first time, a substantial number of students and IS professionals working in China who attended an ETHICOMP working conference held in that country in April 2007. This has enabled a comparison to be made between the responses of each of these disparate groups, and between them and previous survey respondents.

It is not surprising to find that although the students show a high level of awareness of some ethical issues, they are generally less aware than the academics, and there are some areas where the student responses are a cause for concern. There are some intriguing differences within the UK student group. It is interesting to find that the responses of those attending the conference in China are aligned with the UK students' responses on some issues, with the ETHICOMP 2007 delegates" responses on others, but in some cases with neither of these other groups.

It was found that most UK students believed it was unacceptable to make unauthorised copies of commercial software to use at work whilst a quarter of them thought it was acceptable to make unauthorised copies of commercial software for their own private use. By contrast, Chinese students are more likely to find unauthorised use to be acceptable when it is for use at work than when it is for their own private use.

One of the findings indicates that the privacy of data is most valued by the practitioners and the experienced academics, and is less appreciated by the UK and Chinese students. There is a clear warning for organisations that they need to ensure that their privacy and security policies are clear, communicated to all employees, and that employees" awareness and deployment of them are continually reviewed.

In the current climate of reported major system failures it was pleasing to find evidence that education may be having an effect on attitudes to testing. It is hoped this responsible attitude will be carried into the workplace by newly qualified graduates. It remains the case that employing organisations have a responsibility to provide an environment that encourages ethical practice and to ensure that commercial pressures to meet budgetary and other deadlines do not lead to ethically dubious practices such as cutting down on testing. Project leaders have a responsibility not to agree to unrealistic deadlines that could result in such pressures being applied; professional societies have a supportive role to play in helping members to maintain their integrity in the face of pressure from employers.

Overall, the report contains recommendations for organisations, for professional societies and for educators that are intended to promote more socially responsible practices within the IS community. The key points:

Organisations should:

- adopt a Code of Conduct for all employees
- introduce a clear policy concerning the use of computing resources by employees for their own activities, and consider allowing the use for selected non-profit-making activities as a contribution to the local community or as a legitimate perk for employees
- establish clear guidelines for the introduction and operation of any electronic surveillance process, including email and internet usage monitoring, ensuring that all employees are fully consulted and that their rights to privacy in the workplace are respected
- promote an approach to systems development that encourages genuine stakeholder involvement in decision-making
- improve the promotion of data protection awareness among staff and review the means by which compliance with data privacy and data protection requirements are assured
- make greater efforts to provide a working environment that encourages ethical practices, supporting employees in resisting the temptation to allow commercial pressures to lead to ethically dubious practices—instead, promoting their ethical stance to their commercial advantage

Professional societies should:

- ensure that their Code of Conduct remains up-to-date and relevant to the profession, increasing efforts to promote awareness of the Code among members and providing guidance how it can be applied in practice
- provide a greater degree of particular support for their younger members, helping them to acquire greater awareness of the ethical issues they will encounter throughout their careers
- promote members' awareness of the role that IS staff play in the designing of data privacy and data protection compliance into information systems

Educators should:

- address ethical issues more extensively in their curriculum, to raise the awareness of young, aspiring professionals concerning all of the issues covered in this survey
- include in their research agenda the ethical and social impact issues such as access, workplace monitoring, global workforce, professional obligations and choice

7.11 Beyond Technology [2008]*

This section was first published as: Rogerson, S. (2008) ETHIcol—Beyond Technology. IMIS Journal, Vol 18 No 6, pp. 27–28. Copyright © Simon Rogerson

ETHICOMP 2008 had the overall theme of "Living, Working and Learning beyond Technology." In 1995, Rogerson and Bynum wrote, "The information revolution has become a tidal wave that threatens to engulf and change all that humans value. Governments, organisations and individual citizens therefore would make a grave mistake if they view the computer revolution as "merely technological." It is fundamentally social and ethical." This issue is still prevalent today. Indeed, Professor Virginio Cantoni of the University of Pavia writes, "In the era of globalisation, it is essential to develop qualities like adaptability and ease of social integration; receptivity towards others; the ability to observe carefully and discover the facts; awareness of the importance of interpersonal relations; critical self-evaluation of outcomes; and the ability to respect deadlines and decisions." We need to consider ICT in context and as a facilitator of social interaction, human endeavour and environmental wellbeing. We must not simply live with ICT but live beyond it.

With the advent of converging technologies we have become more and more dependent on blogs, chat rooms, wikis, podcasts, and so on to interact with each other socially, educationally and professionally. The technology, which in hindsight was so primitive in the 1970s, the decade heralding the Information Age, has become much less primitive, more flexible and more accessible through, for example, Web 2.0, social network sites, Voice over Internet, 2nd Life, mobile communication, wireless/broadband, on-demand multimedia and embedded computer technology coupled with falling unit costs.

The papers at ETHICOMP 2008 provided a rich dialogue about the impact of this less primitive technology. Three contributions illustrate this.

Gordana Dodig-Crnkovic and Margaryta Anokhina explained that ICT contributes effectively to spreading of all sorts of information, including gossip and rumours. What was interesting was that the old human practice of gossip had taken on a new form in the Information Age. Gossip and rumour resides in public space but the presenters argued that in the Information Age we need to re-think what public space such as working place means in terms of relationships with different degrees of closeness, how interpersonal relationships are configured and implemented and how their architecture shapes their functions and meanings.

The nature of ICT and its impact on accessibility was discussed by Alfreda Dudley. It was argued that to improve access and usage will require significant effort and resources from several sectors: designers, producers, and representatives of citizens and consumers themselves. In order to move closer towards a level playing field, for all individuals in society, with regards to ICT access, it would take the combined efforts of institutional and community entities to make major structural and policy changes.

In discussing the advance of Second Life, Denise Oram vividly illustrated that ICT is still primitive. She explained that the virtual world has become too far removed from the real world. The virtual world is the world of bits, bytes, records and files, whatever fancy graphics, video, search strategies are used to dress it up. The challenge is to put the real world back into the virtual world. Making a character on a video screen react physically to a ball I throw at it or a rock I throw at it is an issue of bits and bytes. Making it react realistically to an emotional speech I throw at it is the real challenge.

The widening engagement of the ICT evolution from the primitive to the less primitive has seen different groups becoming involved for the first time at different points. First the scientists and philosophers focused on concepts and possibilities, seeking out the truth. Then the technologists, who at first were elitist, took these concepts and developed new forms of ICT. Then followed professional business people and industrialists, who started to embrace ICT, in order to solve increasingly complex problems. Gradually ICT seeped into our everyday lives. It was then ICT-knowledgeable citizens enthusiastically consumed ICT, but then often became sceptical because of failing systems and broken technological promises. Consequently, many citizens today tend to be reasonably realistic about ICT's potential and reliability. Finally, when ICT becomes universally acceptable, accessible, useful, usable and trustworthy it will become invisible. Only then will nearly everyone use this technology without question or concern. We will then live and work beyond technology.

"Living Beyond Technology" does not mean living without it. On the contrary, it means using it without even noticing it. It means, relying upon it without even thinking about it. The better the science and the better the technology, the more unnoticed and reliable technology can become and the more "fit for purpose" it can be.

7.12 Direction of Change [2010]*

*This section was first published as: Rogerson, S. (2010) ETHIcol—Direction of Change. IMIS Journal, Vol 20 No 2, pp. 37–38. Copyright © Simon Rogerson

The ETHICOMP conference series was launched in 1995 by the Centre for Computing and Social Responsibility. The purpose of this series is to provide an inclusive forum for discussing the ethical and social issues associated with the development and application of Information and Communication Technology (ICT). Conferences are held about every 18 months. There have been eleven conferences in Europe and beyond. Delegates and speakers from all continents have attended and over 780 papers have been presented. IMIS has been a sponsor of ETHICOMP from the onset.

ETHICOMP 2010 had the overall theme of "The backwards, forwards and sideways changes of ICT." Society has changed dramatically over the last sixty years with the advent of ICT. However, as Alvin Toffler wrote "change is non-linear and can go backwards, forwards and sideways."

Some ICT-related changes will be good and move society forwards, for example, the ability to diagnose hundreds of diseases quickly and accurately, the ability to find lost children and lost pets easily and quickly using a Global Positioning System (GPS) and the ability to help persons with disabilities overcome them or compensate for them.

Some ICT-related changes will be bad and cause harm, for example, the ability to create computer viruses, the ability to detonate bombs from many miles away over the internet or telephone and the ability to widely spread child pornography.

There are some ICT-related changes that appear to have no effect at all simply moving us sideways, for example, the ability to enter numbers into a machine using a keyboard, the ability to make a phone call from your home to someone else's home and the ability to quickly make a list of items.

At ETHICOMP 2010 there were several papers relating to social media. These included; *Social networking and the perception of privacy within the millennial generation* by Andra Gumbus, Frances S. Grodzinsky and Stephen Lilley; *Employing social media as a tool in information systems research* by Michael J. Phythian, N.Ben Fairweather and Richard G. Howley; *The Twitter revolution* by Wade L. Robison; *An examination of the impact of ICT, particularly social networking, on the education and experiences of young people of secondary statutory school age* by Anne Rogerson and *The social and ethical implications connected with the development of social networking websites* by Janusz Wielki. A brief analysis of these five papers against Toffler's maxim reveals a number of issues.

Positive forward changes through social media are manyfold. For example, social media is used beyond socialising to seek advice and professional development as well as offering new business uses. It creates a collective intelligence across society through interactive collaboration across fast communication networks. Citizens are empowered and their voice is heard. This helps in establishing positive relationships such as that between the trader and the consumer. It provides an opportunity to break down barriers and interact across cultures and countries.

There are several harmful backward changes that have resulted from the advent of social media. For example, there are potential losses in privacy. There is increasing profiling of consumers and job applicants from information to be found on social media networks. Social media, particularly when it is unregulated, provides platforms for harming children. Social interaction can become stilted through the use of social media incapable of supporting all aspects of human communication in a flexible and adaptive manner.

Social media has resulted in several sideways changes which have had little effect. For example, even though there exist potential serious threats to personal privacy and abuse of intellectual property rights, there has been little change in people's attitudes to such issues. The cherished goal of universal access remains unfulfilled with social media resulting in little movement in the digital divide. People remain ignorant of the impacts of the speed of change of technology such as social media.

There is a tendency to consider Toffler's maxim in a one directional manner—forward is positive whilst backwards and sideways is negative. But is this really the case? If society is to progress it will have to embrace ICT. Progress is when societal core values are sustained and promoted through fit-for-purpose ICT. Thus, ICT type-positive is where there is appropriate technological functionality and when it is enabling and inclusive. ICT type-negative is where there is increased technological functionality per se and when it is restrictive and exclusive.

Progress in the Information Age is thus multi-directional using Toffler's maxim. Forward change is progress if ICT type-positive is implemented. For example, ICT applications which enable socially excluded people to engage more easily is progress. Backward change is progress if ICT type-negative is retired, removed or replaced. The recent proposal in the UK, by the new coalition government, to remove digital identity cards is a backward change which many consider to be progress. Sideways change can be deemed progress if neutral amoral applications are developed and implemented. Online travel timetable systems which give us the same information but in different forms of access is primarily a sideways change but is progress.

We have become addicted to information and communication technology. Many aspects of society now have a physical, psychological and/or economic dependence on ICT. Such dependence has been fuelled by technologists, businesses and governments with their overly-optimistic predictions of the potential benefit to us all of embarking down the digital pathway. Those who demand caution are often dismissed as sceptics or heretics. In planning to add ICT the ethical dimension must be considered so the potential positive, negative and neutral impacts on society, organisations and individuals can be understood and appropriate action taken.

7.13 IS IT Ethical? 2010 Report [2010]*

*This section was first published as: Rogerson, S. & Prior, M. (2010) ETHIcol—IS IT Ethical? 2010 ETHICOMP Survey of Professional Practice. IMIS Journal, Vol 20 No 4, pp. 35–36. Copyright © Simon Rogerson & Mary Prior.

This edition of ETHIcol focuses on the 2010 Report of the ETHICOMP Survey of Professional Practice which is now available from IMIS. Data was collected during 2009 and subsequently verified and analysed in 2010. It reports the attitudes of Information Systems (IS) professionals and entrant IS professionals to a variety of ethical issues.

This is the sixth report in the series which began in 1998 and there has been an increase in the overall number of respondents to 329 in total, representing a range of age, experience, type of employing organisation and country of employment. For some issues, these factors appear to have some influence on respondents" attitudes.

Here are some highlights from the report.

The tendency perceived in earlier surveys for younger, less experienced respondents to show less ethical awareness in some areas than their more experienced colleagues, is confirmed. There are differences in response between respondents working for some types of employing organisations and in certain geographical areas.

There remain a number of areas where more discussion during the education and preparation for professional practice is required, and where more guidance and support for individuals in the workplace is desirable to encourage consistently responsible behaviour.

From the responses to the survey, a picture emerges of young professionals who care about the overall objectives or purpose of the projects they work on, but who may not have the freedom to refuse to work on any that they consider to be unethical. The more mature professional in senior management posts is more able to be more discriminating about his or her work.

The vast majority of self-employed or public service workers are most likely to disagree about employees being able to re-create intellectual property for an employer other than the one who originally paid for it. There is still a slight majority of respondents from private enterprise who disagree, but the figures are smaller. However, when it comes to respondents working for academic organisations, attitudes are evenly split between disagreement and agreement.

Respondents working for private enterprise of all types, and in public service, are fairly evenly split on the acceptability of using their employer's computing facilities for *non-profit* use, whereas among those working for academic institutions the level of agreement is much higher. This may be because of the difference in work environments with academia blurring the boundary between work and home more than in other organisations.

There is some variation between different countries to the statement, "*The licensing of computer professionals should be introduced into my country.*" The highest level of support comes from African respondents. The lowest level of support for licensing is found among respondents from the US, with UK respondents falling between them and the rest of Europe/ Australia.

The report concludes with a series of recommendations to organisations and professional societies. It is concerning that several of these have appeared in previous reports suggesting that in some areas change of attitudes and improvement in practice are long overdue.

Organisations are recommended to:

- seriously consider adopting a Code of Conduct for all employees
- increase efforts to promote awareness among all employees of:
  - ethical issues
  - the organisation's Code of Conduct
  - how the organisation's Code of Conduct may be applied to guide ethical decision-making
- establish "whistleblowing" procedures to encourage employees who become aware of unethical practices within the organisation to come forward
- introduce a clear policy concerning the use of computing resources by employees for their own activities, and consider allowing the use for selected non-profit-making activities as a contribution to the local community or as a legitimate perk for employees
- establish clear guidelines for the introduction and operation of any electronic surveillance process, including email and internet usage monitoring, ensuring that all employees are fully consulted and that their rights to privacy in the workplace are respected
- review on a regular basis the security of computer-held data with attention to both technical aspects and management aspects affecting potential threats
- consider and clarify their policy concerning the re-creation of intellectual property such as a product, program or design by employees when they move to another employer
- ensure that all policies are clearly communicated to employees, in particular to student and new graduate employees, and are deployed throughout the organisation

- promote an approach to systems development that encourages genuine stakeholder involvement in decision-making
- improve the promotion of data protection awareness among staff and review the means by which compliance with data privacy and data protection requirements are assured
- make greater efforts to provide a working environment that encourages ethical practices, supporting employees in resisting the temptation to allow commercial pressures to lead to ethically dubious practices—instead, promoting their ethical stance to their commercial advantage

Professional societies representing the IS profession are recommended to:

- ensure that their Code of Conduct remains up-to-date and relevant to the profession, increasing efforts to promote awareness of the Code among members and providing guidance how it can be applied in practice
- provide a greater degree of particular support for their younger members, helping them to acquire greater awareness of the ethical issues they will encounter throughout their careers
- promote debate of the continuing applicability of legislation such as the software licensing laws, in the light of current developments, opinion and practice
- promote debate concerning the desirability of licensing for information systems professionals
- promote members" awareness of the role that IS staff play in the designing of data privacy and data protection compliance into information systems

7.14 Social Impact of Social Networks [2011]*

*This section was first published as: Rogerson, S. (2011) ETHIcol—Social Impact of Social Networks. IMIS Journal, Vol 21 No 4, pp. 28–29. Copyright © Simon Rogerson

IMIS has been a sponsor of the ETHICOMP conference series since the series started in 1995. The purpose of ETHICOMP is to provide an inclusive forum for discussing the ethical and social issues associated with the development and application of Information and Communication Technology (ICT). In September 2011 the 12th conference was held at Sheffield Hallam University, UK and the theme was *The Social Impact of Social Computing*. Delegates attended from around the world to debate this important topic.

Social computing, or social media as it is often called, is a group of internet-based applications that build on the ideological and technological foundations of Web 2.0, and that allow the creation and exchange of user generated content. Kaplan and Haenlein (2010) explain there are two elements to social media; media-related dimension and social dimension. Media-related refers to social presence, be it acoustic, visual or physical and media richness which is about the amount of information transmitted in given time intervals. The social dimension refers to self-presentation, which is about people's desire to control the impressions other people form of them and to self-disclosure which is about the conscious or unconscious revelation of personal information that is consistent with the image one would like to portray.

Many papers at ETHICOMP 2011 addressed these concepts. There were several about Facebook and Twitter. Miranda Mowbray in her paper *A Rice Cooker Wants to be My Friend on Twitter* explained that everyday objects that tweet include toasters, garage doors, shoes, ovens, houses and cat-flaps. This activity by non-human agents adds a strange twist to the online social world. Don Gotterbarn in his paper *Tweeting is a beautiful sound, but not in my backyard: Employer rights and the ethical issues of a tweet-free environment for business* focused on the tensions between the social world and work. He explained that corporate policies and restrictions on computer use in the workplace used to be justified on the ground that the computers were owned by the corporation. Now employees bring their own computers to work in the form of, for example, smart phones, iPods and iPads. In addition, social networks have blurred the distinction between personal and corporate information. Ryoko Asai in her paper *Social Media as a Tool for Change* compared Tunisia and Japan. she explained that people have two essential and indispensable places in their lives: one is the home and another is the work place. In addition to those two places, people usually have another place where they can have social relationships with others informally. The third place nurtures relationships with others and mutual trust. Social media has created a new third place and people from different cultures such as in Tunisia and Japan react in different ways. Malik Aleem Ahmed in his paper *Online buzzing—using friendship for benefits* discussed the influence on and the direction of people by their friends. He explained that firms have developed different ways to market their products and services on the internet using people's friendships. He discussed the moral issues that arise when people and businesses use social media to get friends to do something or to influence their political views.

The impact of social media was vividly highlighted by the August riots in the UK. No one riots alone. Riots are only possible when large numbers of people both share and see each other as sharing the same antagonisms and motivations. So did social media play a role in the August riots? There us much evidence in the public domain to show services such as Facebook, Twitter and Blackberry Messenger were used by rioters to inform others of what was happening and in some instances encourage people to go onto the streets. There was a call to place restrictions on social media because of such usage. However, those who claimed social media had caused the riots were wrong. Social media as a technology was used as a tool, but the common motivations had to exist in the first place. It is interesting that social media as the technological enabler was used in a positive way as well. For example, the police used social media to track down rioters and looters as well as send out counter-communications in an attempt to calm the situation. There were many instances of community groups using social media to organise clean-up operations and to reclaim their communities.

Writing in the Guardian on 24 August, Dan Sabbagh raised a series of difficult questions about social media. These included a number of ethical questions.

- Should Facebook, Twitter, and Blackberry Messenger be restricted or shut down in times of civil disorder?
- Can all three networks, and in particular the private BlackBerry Messenger, do more to help the police?
- Should Facebook, Twitter, and RIM give the police privileged access to their networks at time of civil disorder, or help law enforcement by providing more real-time data analysis that could help the police prevent rioting?
- Should mobile phone companies monitor phone calls in a riot situation—or be prepared to scan for inflammatory text messages sent over their networks?
- Should people who incite violent disorder be subject to four year jail terms, or similar stiff sentences, as were handed down by Chester Crown Court to two men who posted messages on Facebook encouraging other people to riot in their home towns?

Alex Buttle, mobile phone and social media expert at uSwitch.com stated:

What's become clear is that while social media can be a powerful force for good and for fun, it can also be subverted for more sinister purposes. Via social media platforms such as Twitter, Facebook and BlackBerry Messenger, people can mobilise and influence other like-minded people in real time and all from the mobile phone in their pockets. Social media may still be in its infancy, but the London riots have shown us its immense power.

What is disappointing about the discussions on social computing (and ICT in general) is that it takes a catastrophic event, such as widespread rioting, to as a catalyst for discussion, often in academia, about the potential problems to be turned into action by industry and government. So, what is needed is a comprehensive analysis of social media to understand fully the social impacts of this powerful technology. This must be presented in a meaningful way to industry and government so that informed action can occur. Only then can effective measures be devised and implemented that enables us to use social computing in a positive way, whilst ensuring the negative uses are restricted.

7.15 An Ethics Progress Litmus Test [2015]*

This section was first published as: Rogerson, S. (2015) An Ethics Progress Litmus Test, ITNOW, Vol 57, No 4, p. 62. Copyright © Simon Rogerson. DOI 10.1093/itnow/bwv113

BCS, the Chartered Institute for IT, ran a special edition of *ITNOW* in the autumn of 2014 which focused on Ethics in ICT. In many ways it is a litmus test of ethics progress by academics and practitioners working in tandem. It is a disappointing read.

Runciman (2014) points out the "philosophical challenges of extraordinary complexity" in the US Navy's pursuit of embedding moral competence in computational architecture of warfare technology. This project smacks of arrogant technological determinism which is so dangerous. The discussion by Bennett (2014) on robot identity assurance concludes with a series of uninspiring recycled actions. For example, "debate about the use of RFID and NFC technologies which enable tracking of individuals without their knowledge or consent." was an action called for many years ago (for example, see Rogerson (2004a)).

Southey (2014) discusses the every-increasing scope of ICT application. He concludes, "The ethical dilemma that faces us is therefore: can I justify unleashing this IT development, knowing that I do not know the extent of its safety? Have I even come close to imagining the worst that could happen? Of course, we can argue, the IT profession is not regulated like law or medicine; BCS has a voice but, unfortunately, no real clout. If we refused to work on robot soldiers, someone else will do it." Once again he simply restates observations of the past. The same is true of Dainow (2014) who discusses the ethics of emerging technology. He concludes, "The IT professional is moving to join the doctor at the centre of modern ethical concerns for society. Society's gaze is sure to follow. It is no longer viable for IT professionals to remain ethically neutral. The next generation of technology will inevitably generate more controversy and concern than anything seen so far. We have enough experience to anticipate many of the issues and avoid them through conscientious and ethically aware design."

Cultural diversity is explored by Freeland (2014). He concludes by highlighting gender discrimination as a key issue in ICT application. This has been known about and investigated for over 20 years. Freeland's article seems naïve and shallow. Holt (2014) focusses on the issue of ethically fit-for-purpose. Once again her conclusions are disappointingly lacking in new insight when she writes as IT professionals we have a duty to build a mind-set of considering the wider consequences of the IT solutions our developers design . . . we need to contribute to the wider debate of how IT solutions are used, and how ethical decisions are made around IT-enabled concepts. . . . So finally, let's get our professional bodies involved in leading the way to develop policy and opinion pieces before our politicians enforce laws, or our judges pronounce life-changing judgments that result in even greater ethical issues.

Twenty years ago I wrote (Rogerson, 1995),

no longer can the profession seek absolution through focusing only on the technical agenda. Indeed, the first question any IS/IT professional should ask is "Is the action ethical?" and be able to answer based on reasoned thought.—We all need to act and act now!

It is disheartening to find Holt writing similar statements in 2014.

Overall, it is a disappointing edition of *ITNOW*. The lack of ethical consideration in systems design and implementation is evident. The calls for action are neither new nor inspiring. There is virtually no evidence and no pragmatic action; the emphasis being on top-down political rhetoric. In many ways this edition illustrates at best that we have stood still but probably we are moving backwards in the quest for ethically-acceptable technological implementations. There is little evidence of drawing from more than 20 years of effort in developing ICT ethics thinking and practical approaches. Even more surprising is that there is no mention or use of past BCS efforts in addressing ethics (for example, see Harris et al. (2011)).

In 1995, Terry Bynum and I wrote (Rogerson & Bynum, 1995),

The brave new world of the Information Society—with its robots and global nets, telemedicine and teleworking, interactive multimedia and virtual reality—will inevitably generate a wide variety of social, political, and ethical questions. What will happen to human relationships and the community when most human activities are carried on in cyberspace from one's home? Whose laws will apply in cyberspace when hundreds of countries are incorporated into the global network? Will the poor be disenfranchised—cut off from job opportunities, education, entertainment, medical care, shopping, voting—because they cannot afford a connection to the global information network? These and many more questions urgently need the attention of governments, businesses, educational institutions, public advocates and private individuals. We ignore ethics and computing at our peril.

The evidence from the *ITNOW* special edition suggests our warnings are yet to be heeded.

7.16 Computing by Everyone for Everyone [2019]*

*This section was first published as: Rogerson, S. (2019) Computing by everyone for everyone. Journal of Information, Communication and Ethics in Society. Vol 17 No 4, pp. 373–374. Copyright © Emerald Publishing Limited. Reprinted by permission. DOI 10.1108/JICES-11–2019–098

Computing is no longer the sole domain of professionals, educated and trained through traditional routes to service public and private sector organisations under paid contracts. Computing has been democratised with the advent of economically accessible hardware, a multitude of software tools and the internet. Computing is by everyone for everyone. The youngest app developer at AppleWorldwide Developers Conference in June 2019 was Ayush Kumar, a 10-year-old boy who started coding when he was 4.

Much effort has been expended in the creation and dissemination of excellent codes of ethics by many professional bodies allied to computing. The adoption of, adherence to and effectiveness of such codes in practical computing have to be questioned with the continued occurrence of so many system failures and also illegal activities that occur leading to public outcry. Even though the codes of ethics exist, why is it that significant unethical activity within computing remains? IT development is a global activity about which IDC periodically produces surveys. The IDC survey of 2018 found that there were, worldwide, 18,000,000 professional software developers and 4,300,000 additional hobbyists (www.idc.com/getdoc.jsp?containerId=US44363318. accessed July 2019). The combined membership of leading professional bodies, ACM, ACS, BCS, and IFIP (assuming no joint memberships which in practice will not be the case), represents only 3.09 per cent of that global total. This suggests that, on the basis of statistics, professional bodies allied to computing and their adopted codes of ethics have little influence on practical computing. Of course, this argument does not take into account the nature and importance of the systems developed and who is developing them. Nevertheless, from these statistics, it is clear that a large global population needs to engage in a new form of dialogue regarding the ethics of practical computing. This might include, for example, accessible exemplars of good and bad practice, interactive case analyses of failed systems and a universal charter for computing, which would be the foundation of computing education from the start of a child's education through to becoming a computing practitioner. Codes of ethics are important, as they provide the detail on which sound computing strategies can be planned and implemented. However, to suggest, these alone can be used to resolve unethical computing practice is folly. A new approach which engages all members of society is needed. Why? Because society is now computing dependent and anyone can develop computing systems which might be used by thousands, if not millions, of people. Impacts, whether positive or negative, spread rapidly and are very difficult to reverse.

IT application, research, innovation, availability, and use are affected by three types of drivers. There are top-down drivers which are typically impositions by bodies of authority which dictate where resources should be placed to achieve some overall goal. Bottom-up drivers emanate typically from grassroots collective action resulting in a widespread change. Middle-out drivers involve all those within, for example an organisation, who are empowered to initiate change, support it, propose new ideas and innovate. These drivers affect attitudes and societal norms. Indeed, the amalgam of top-down, middle-out and bottom-up drivers leads to a complex situation where the attitude and behaviour of individuals and collectives involved in IT are highly influential in the delivery of socially acceptable IT. Therefore, with each passing day, information ethics becomes more important, as it is that which steers in an ethical direction for all those involved in IT.

Millennials and post millennials, who now represent around half of the global workforce, are key bottom-up drivers because they, as citizens, have grown up with technology and consider change as ever-present. Technology is their sixth sense. Increasingly millennials will influence the way in which society looks at IT, what is acceptable IT and what is not. The demand for more flexible working and the blurring of traditional boundaries between home and work will increase. The millennial voice must be heard and must be taken into account, for they are the future. Their information ethics education and sensitivity must be nurtured to promote socially acceptable IT.

Without practical information ethics, society is increasingly at risk from many threats. Here are just three examples.

The risk of dual use has always existed and, at times, has occurred. Free and open-source software (FOSS), like most technology, is dual use technology; it is a technology which can be used for both "good" and "evil" purposes. The enormous range of powerful FOSS software available to anyone for any purpose greatly increases dual use risks to society.

More than 50 per cent of reported crimes are related to the online world. For the criminal, cybercrime remains low risk with high return. There exist criminal alliances through which cybercrime information is shared using, for example, the Dark Web. Cybercrime is systemic and global in nature. Phishing emails and associated malware remain the greatest threat, but ransomware is rapidly increasing and denial of service attacks are resurgent. Cybersecurity strategies and associated actions are needed to reduce the effects of cybercrime. However, this cannot simply be a technological endeavour. Education should be a key element. This includes educating teenagers who can be lured by the perceived excitement of cybercrime, making parents aware of the dangers facing their children in the online world and training employees to identify and resist the temptation of clicking links in phishing emails.

Within industry and government, the compliance culture has taken a firm hold and so strangles the opportunity for dialogue and analysis of complex multi-faceted socio-ethical issues related to IT. Superficial compliance is dangerously unethical and must be challenged vigorously in a technologically dependent world. The time frames for IT development and IT regulation and governance are, and will always be, misaligned. By the time some control mechanism is agreed, the technology will have moved on several generations and thus what has been agreed is likely to be ineffective. This seems to be the case with the latest trend, Artificial Intelligence (AI) Governance, as there are so many opinions and vested interests causing protracted debate while AI marches onwards. Thus, it is paramount to imbue strategists, developers, operators and users with practical information ethics. In this way, ethical computing has a chance of becoming the norm.

To conclude, all those involved in computing, albeit as provider or consumer, need to have the ethical tools, skills and confidence to identify, articulate and resist unethical aspects of IT. Moreover, they should be free to challenge the decisions of and orders issued by IT leaders where those actions are ethically questionable, without detrimental effect to themselves.

7.17 Digital Ethics is The Paradigm Shift [2020]*

This section was based on: Rogerson, S. (2020) Digital Ethics is The Paradigm Shift. This was part of the online opening ETHICOMP 2020 keynote discussion, A journey in Computer Ethics: from the past to the present. Looking back to the future, 16 June 15.30–16.00 (Madrid time) by Simon Rogerson (De Montfort University, UK), Shalini Kesar (Southern Utah University, US), Don Gotterbarn (East Tennessee State University, US), Katleen Gabriels (Maastricht University, Netherlands). Copyright © Simon Rogerson.

1995, the year that Apple launched its first WWW server, Quick Time On-line; the year Microsoft released Internet Explorer and sold 7 million copies of Windows 95 in just 2 months.

On 28 March 1995 I opened the first ETHICOMP. I said, and I quote,

> We live in a turbulent society where there is social, political, economic and technological turbulence it is causing a vast amount of restructuring within all these organizations which impacts on individuals, which impacts on the way departments are set up, organizational hierarchies, job content, span of control, social interaction and so on and so

forth. With that list there is a lot that we can be talking about in the next three days and the next three decades.

That could be today! Little did I realise that 25 years on I would be talking about the very same things.

In 1995 I explained that

> Information is very much the fuel of modern technological change. Almost anything now can be represented by the technology and transported to somewhere else. It's a situation where the more information a computer can process the more of the world it can actually turn into information. That may well be very exciting but it is also very concerning.

In May this year Mobile App Daily published its 2020 technology trend forecast—It demonstrates the world is now digitised through, for example, 5G, clouds, AI, algorithms, augmentation, machine autonomy, data analytics, edge computing and the Internet of Things. This digitization of everything requires a greater emphasis on what we should now call Digital Ethics. If not, then a very bleak, discriminatory world beckons.

Let me give you just one example which relates to my 1995 comment,

> we have a technology which is very impressive but it's technology the problem. A lot of things it claims to do don't actually happen in practice. It does actually drive us apart rather than bring us together in lots of instances. It has made the world move faster and I don't know whether we actually want to move faster. It fragments us and it [*the technology*] doesn't [*care*] matter about who it leaves behind.

The world turned to the online with the advent of COVID-19. The URL is ubiquitous in the fight against the pandemic. Digital natives are now informed, socialise, shop, learn and are entertained in a post pandemic virtual world. But the digital outcasts cannot access this virtual world because of lack of opportunity, lack of finance, lack of wherewithal, digital aversion. Little attention has been given to the plight of the digital outcasts—they are the new underclass who, apparently, do not matter. Digital Ethics is needed to address this iniquitous situation of a damaging digital divide.

Digital Ethics is The Paradigm Shift from Computer Ethics, Information Ethics and ICT ethics. Its approach and perspective must be holistic and synergistic. Technology is now done by everyone for everyone. Silo mentalities of professional bodies such as BCS and ACM, and academia such as Harvard, which continue to focus on minority computer science populations, must be challenged. They need to be replaced by inclusivity and empathy that will embrace a global community from the cradle to the grave. In that way fit-for-purpose technology for everyone is more likely.

7.18 Rebooting Ethics Education in the Digital Age [2021]*

*This section was first published as: Rogerson, S. (2021). Rebooting ethics education in the digital age. Academia Letters, *Article 146. Copyright © Simon Rogerson. DOI 10.20935/AL146.*

1951 was the year that J. Lyons & Co. used a computer (LEO) to manage stock for its renowned chain of over 200 high-street cafes and tea shops. It was the first time across the world that computers were used commercially for business data processing (Ferry, 2003) and marked a sea change. 1995 was the year that Apple launched its first WWW server, Quick Time On-line. It was the year Microsoft released Internet Explorer and sold 7 million copies of Windows 95 in just 2 months. This was a second global sea change; the move away from computers and data processing to the digital age via the information superhighway. It was a move away from highly educated, usually, men in lab coats undertaking elitist science and technology, shifting to a world where almost everybody uses accessible digital technology which could have been developed by almost anyone including young children and elderly adults.

In the digital age it is people who change things. It is people who make digital technology. It is people who use and abuse digital technology. The tension between use and abuse is where the ethical hotspots lie. Digital technology can add value to life, but it can also take value away from life. Some ethical hotspots may be obvious whilst others may not. All must be addressed so that the digital age is good for everyone as well as for the world at large. This can only be achieved through effective digital ethics education and awareness programmes. Such programmes should promote social responsibility which should aim to (Rogerson, 2004b):

- Develop a socially responsible culture within work, home and society which nurtures moral individual action
- Consider and support the wellbeing of all stakeholders
- Account for global common values and local cultural differences
- Recognise social responsibility is beyond legal compliance and effective fiscal management
- Ensure all processes are considered from a social responsibility perspective
- Be proactive rather than reactive when addressing the development and use of digital technology

Digital ethics education and awareness must develop the individual's confidence and skills, through lifelong learning, and so provide the tools to enable everyone to act responsibly and ethically. Discussion, dialogue, storytelling, case study analysis, mentoring, and counselling are examples of techniques that can be used to nurture practical wisdom and insight which will lead to virtuous citizens of the digital age.

Digital ethics education needs to be rebooted because the current effort seems to be too narrowly focused and so is probably ineffective. For example, Grosz et al. (2019) discuss a recent initiative by Harvard to embed ethics in the computer science degree curriculum. The initiative has been heralded by many, but it is problematic in that it simply reinvents the wheel of computer ethics education which has a long and comprehensive history stretching back to the 1980s. It offers little new insight and does not link to a 40-year history and experience. It misaligns with the demographic profile of the computing community in the digital age and fails to acknowledge that digital ethics education in the post millennial era is best started from early childhood. Similarly, this misalignment seems to be the case with the proposals by Blundell (2020) in the December 2020 edition of ITNOW, the membership magazine of BCS, The Chartered Institute for IT, which featured lifelong learning.

As mentioned in Section 7.16, the IDC 2018 survey found that there were 18 million professional software developers and 4.3 million hobbyists. According to the published membership figures of the leading computing professional bodies, only 3% of the total 22.3 million belong to a professional body. This suggests that professional bodies, in their current role, have little influence on 97 percent of global software developers whose moral code and attitude to social responsibility comes from elsewhere.

It is not that ongoing digital ethics training as part of professional development programmes is wrong; it is that this should not be the only or dominant element in the digital ethics education landscape. Those entering the computing profession are faced with a plethora of application areas using a vast digital technology armoury. For new entrants, the responsibilities and obligations to society are onerous. Yet it is uncertain how well they are prepared for such challenges and whether they have been educated to understand that they are the custodians of the most powerful and flexible technology humankind has invented. Too many computing professionals hide in technological clouds seemingly indifferent to the ethically charged nature of digital technology. It is unclear whether this is through lack of awareness or a belief such issues are outside their scope of responsibility. The VW emissions scandal is a prime example (Rogerson, 2018).

Digital ethics education for everyone should start at an early age and in that way the foundations are laid for responsible adulthood in the digital age. This is a challenging proposition for as Churchland (1996) explains, the development of moral character in children takes time. The formality of school, and the informality of home and social settings are equally important, particularly in the early stages of this journey to moral maturity. It is a journey which starts the moment a child is born, continues through childhood into adolescence and finally into adulthood. Novel cross-discipline educational experiences should be developed which capture the interest and imagination of learners thereby firmly establishing digital ethics as a pillar of their conduct and their participation in the digital age. Some possible avenues to explore are (Rogerson, 2020a):

- Interactive digital ethics exhibitions in science and technology museums
- Investigating social impacts across the history of computing
- Using thought experiments to explore dual use dilemmas in digital technology
- Using poetry to understand the breadth and depth of digital ethics

Such avenues also offer greater awareness to the public at large and align with Burton et al. (2018), who use science fiction to teach digital ethics.

Rogerson (2020a) discusses the four possible avenues in some detail, however, the last of these avenues, poetry, has been explored in more detail (Rogerson, 2020b). Poetry challenges us to think beyond the obvious and reflect on what has been, what is and what might be. Poetry can reboot the way in which social impact education is delivered to technologists. Rule et al. (2004) explain that incorporating poetry in science and technology teaching expands the curriculum beyond subject knowledge and process skills. Poems can provide meaningful context. Such context is imperative in digital ethics education and awareness at all levels for all people, enabling understanding of the social impacts of digital technology. In partnership, computer science and liberal arts educators could offer an exciting new perspective through poetry as an instrument of presentation and discussion as well as in creative exercises for students of all ages.

It would be appropriate to underpin the digital ethics education reboot with a global beacon. This could be a global digital technology charter which aligned to, for example, The Universal Declaration of Human Rights. Through adapting the work of Franklin (1999) and Barratt (2007), a simple charter has been devised (Rogerson, 2015) which encourages everyone to:

- promote social justice and social care
- restore reciprocity for everyone
- benefit the many rather than the few
- put people first rather than digital technology
- limit economic gain to minimise potential social and environmental cost
- favour the reversible over the irreversible
- create a more inclusive society by reducing barriers and creating new opportunities

We have to accept and adjust to the fact that we are all technologist to a lesser or greater degree. How we educate our future generations must reflect this change to ensure digital technology is societally beneficial. The time has come to reboot ethics education in the digital age.

References

Barratt, J. (2007). Design for an ageing society. *Gerontechnology*, *6*(4), 188–189.

Bennett, L. (2014). Robot identity assurance. *ITNOW*, *56*(3), 10–11.

Blundell, B. G. (2020). Ethics and lifelong learning. *Itnow*, *62*(4), 22–23.

Burton, E., Goldsmith, J., & Mattei, N. (2018). How to teach computer ethics through science fiction. *Communications of the ACM*, *61*(8), 54–64.

Churchland, P. M. (1996). The neural representation of the social world. In L. May, M. Friedman, & A. Clark (Eds.), *Mind and morals: Essays on cognitive science and ethics* (pp. 91–108). The MIT Press.

Dainow, B. (2014). Ethics in emerging technology. *ITNOW*, *56*(3), 16–18.

Ferry, G. (2003). *A computer called LEO: Lyons teashops and the world's first office computer*. Fourth Estate. ISBN: 978-1-84115-185-4.

Franklin, U. M. (1999). *The real world of technology*. House of Anansi Press.

Freeland, A. (2014). Cultural differences. *ITNOW*, *56*(3), 24–25.

Grosz, B. J., Grant, D. G., Vredenburgh, K., Behrends, J., Hu, L., Simmons, A., & Waldo, J. (2019). Embedded EthiCS: integrating ethics across CS education. *Communications of the ACM*, *62*(8), 54–61.

Harris, I., Jennings, R. C., Pullinger, D., Rogerson, S., & Duquenoy, P. (2011). Ethical assessment of new technologies: A meta-methodology. *Journal of Information, Communication & Ethics in Society*, *9*(1), 49–64.

Holt, A. (2014). Flight control. *ITNOW*, *56*(3), 26–27.

Kaplan, A. M., & Haenlein, M. (2010). Users of the world, unite! The challenges and opportunities of social media. *Business Horizons*, *53*(1), 59–68.

Nicolle, L. (1999). The bulletin interview: Simon Rogerson. *The Computer Bulletin*, *41*(4), 18–19.

Rogerson, S. (1995). But IS IT ethical? *IDPM Journal*, *5*(1), 14–15.

Rogerson, S. (2004a). ETHIcol—Tag ethics. *IMIS Journal*, *14*(5), 31–32.

Rogerson, S. (2004b). Aspects of social responsibility in the information society. In G. I. Doukidis, N. Mylonopoulos, & N. Pouloudi (Eds.), *Social and economic transformation in the digital era* (pp. 31–46). IGI Global.

Rogerson, S. (2015, April 7). *ICT codes of ethics*. Keynote Presentation, Ethics Conference 2015, Council of European Professional Informatics Societies (CEPIS), The Hague, The Netherlands.

Rogerson, S. (2018). Ethics omission increases gases emission: A look in the rearview mirror at Volkswagen software engineering. *Communications of the ACM*, *61*(3), 30–32.

Rogerson, S. (2020a). Start a revolution in your head! The rebirth of ICT ethics education. In *Societal challenges in the smart society*, ETHICOMP Book Series (pp. 153–164). Universidad de La Rioja.

Rogerson, S. (2020b). Poetical potentials: The value of poems in social impact education. *ACM Inroads*, *11*(1), 30–32.

Rogerson, S. (2020c). Re-imagining the digital age through digital ethics, invited position paper. In *Promoting character education as part of a holistic approach to re-imagining the digital age: Ethics and the internet webinar* (pp. 25–28). The Jubilee Centre for Character and Virtues, University of Birmingham. Retrieved November 13, 2021, from www.jubileecentre.ac.uk/userfiles/jubileecentre/pdf/news/Cyber-PhronesisWebinarPapers_03(2).pdf.

Rogerson, S., & Bynum, T. W. (1995). Cyberspace: The ethical frontier. *The Times Higher Education Supplement*, *1179*(9), 4.

Rule, A. C., Carnicelli, L. A., & Kane, S. S. (2004). Using poetry to teach about minerals in earth science class. *Journal of Geoscience Education*, *52*(1), 10–14.

Runciman, B. (2014). Drones, robots, ethical decision dilemmas. *ITNOW*, *56*(3), 6–9.

Southey, D. (2014). What could happen? *ITNOW*, *56*(3), 14–15.

UN General Assembly. (1948). Universal declaration of human rights. *UN General Assembly*, *302*(2), 14–25.

Afterword

In 1998, I was commissioned by the Institute of Business Ethics to write a guide for senior executives which explained how to embed digital ethics into corporate strategy and management (see Section 5.3). The guide concludes with (Rogerson, 1998, p. 33):

> Powerful technologies always have important social and ethical implications. IT is the most powerful and most flexible technology ever devised. It can be shaped and moulded to do any activity that can be represented as inputs into logical operations resulting in outputs. The social impact of this powerful technology is growing at an increasing rate. Computers are changing the nature and location of work and indeed the whole way in which society is conducted. The IT revolution is not merely technological it is fundamentally social and ethical.
>
> Organisations must ensure issues such as privacy, ownership, integrity of information, human interaction and community are properly considered.
>
> Computer professionals must be trained so that they are sensitive to the power of the technology and act in a responsible and accountable manner. The adoption of a broader approach that addresses economic, technological, legal, social and ethical concerns will help to harness technology for people's benefit, rather than allowing it to enslave or debilitate them

This still remains true in 2021. Organisations need to adopt this broader perspective to ensure that ethical digital technology flourishes. In 2021, I was invited by the current director of the Centre for Computing and Social Responsibility (CCSR), Professor Bernd Stahl, to write an account of the foundation and development of CCSR as part of the activities which celebrated 25 years of its existence. It was an opportunity to reflect upon how CCSR had become a global leader in addressing the ethical and social issues surrounding digital technology.

For an organisation to mature and expand it needs to function across a range of fronts; CCSR was no exception. A typological analysis was devised to investigate CCSR's journey towards international research standing. Given (2008) explains that typological analysis centres on the development of a set of related but distinct categories which, as a whole, describe an object, subject or phenomenon. For this study, a top-down typological analysis was appropriate which MacNeil (2000) explains uses a bounded set of predefined categories. This set is then populated with the collected data. This structured collection offers a rich description of the entity under investigation. Therefore, a *Presence Typology* was constructed (Rogerson, 2021) which comprised 11 spheres

DOI: 10.1201/9781003309079-8

within which a higher education department or centre needs to have a presence to progress and mature. The 11 spheres are as follows:

1. Academia—The community concerned with the pursuit of research, education, and scholarship sustained by access, openness, inclusivity and freedom.
2. Research—This involves the participation in extending the body of knowledge through original thought, experimentation and application.
3. Teaching—Delivery of appropriate teaching and learning which embraces and explores development and application.
4. Physical—Identity and culture are portrayed through explicit physical entities such as location and layout of accommodation.
5. Virtual—Global visibility is achieved via the virtual world through, for example, social media, video streaming and web portals.
6. Political—Local, national and international political systems will significantly influence the impact of applied research.
7. Industrial—Effective applied research is reliant upon good links with industry and subsequent partnerships.
8. Professional—Engagement is essential with professional bodies which represent, influence, and govern professional practice.
9. Institutional—Participation in institutional leadership, administration, review, and growth.
10. Funding—This covers both the recognised funding bodies and opportunistic funding activities
11. International—Global collaborations with individuals at all levels, as well as collective organisations.

This Presence Typology is specific to Higher Education Departments. However, it is possible to generalise the typology, so it is relevant to any organisation and its departments. This provides the wherewithal for, as I mentioned in the introduction of Chapter 7, "organisations to change the way they look at the feasibility of their activities." The 11 spheres in this *Universal Presence Typology* become:

1. Sector—The community concerned with the pursuit of the particular product or service. The community is best sustained by access, openness, inclusivity and freedom.
2. Research and Development—This involves the participation in extending the body of knowledge through original thought, experimentation and application.
3. Service and Product Delivery—Delivery of useful products and services which are grounded in valid research and development.
4. Physical—Identity and culture are portrayed through explicit physical entities such as location and layout of accommodation.
5. Virtual—Global visibility is achieved via the virtual world through, for example, social media, video streaming and web portals.
6. Political—Local, national and international political systems significantly influence organisational wellbeing.
7. Partnerships—Effective service and product delivery is reliant upon good links and subsequent partnerships both formal and informal.
8. Professional—Engagement is essential with professional bodies which represent, influence and govern professional practice.

9. Institutional and Corporate—Participation in strategic leadership, administration, review and growth.
10. Finance and Funding—Financial governance relies upon effective funding which covers both traditional funding streams and opportunistic funding activities such as venture capital.
11. International—Global collaborations with individuals at all levels, as well as collective organisations.

Involvement and operation across these 11 spheres, taking into account the ethical and social perspectives of digital technology, could thus prove a blueprint and the key to unlocking *Ethical Digital Technology in Practice*.

References

Given, L. M. (2008). *The SAGE encyclopedia of qualitative research methods* (Vols. 1–0). SAGE Publications, Inc. https://doi.org/10.4135/9781412963909

MacNeil, H. (2000). Providing grounds for trust: Developing conceptual requirements for the long-term preservation of authentic electronic records. *Archivaria. 50*, 52–78.

Rogerson, S. (1998). *Ethical aspects of information technology: Issues for senior executives*. The Institute of Business Ethics, ISBN 0952402041.

Rogerson, S. (2021). *The genesis of the centre for computing & social responsibility (CCSR)*. 25 years of the CCSR, De Montfort University. Retrieved October 24, 2021, from www.ccsr.uk/2021/02/25/the-genesis-of-the-centre-for-computing-social-responsibility-ccsr/.

Index

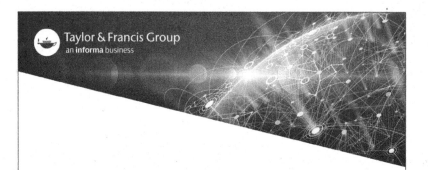

Printed in the United States
by Baker & Taylor Publisher Services

Printed in the United States
by Baker & Taylor Publisher Services